PHILIP E

Weatherwise

The Sunday Telegraph

Companion to the British Weather

MACMILLAN

First published 1995 by Macmillan Reference Books
a division of Macmillan Publishers Ltd
Cavaye Place London SW10 9PG
and Basingstoke

Associated companies throughout the world

ISBN 0 333 61610 3

1 3 5 7 9 8 6 4 2

A CIP catalogue record for this book is available from
the British Library

Typeset by Parker Typesetting Service, Leicester
Printed and bound by Cox & Wyman Ltd,
Reading, Berks

To Mike C and Graham P
who initiated my relationship with
the *Sunday Telegraph*,
and to Harry C
who acted as midwife
to this book

CONTENTS

PREFACE

We are supposed to be a weather-obsessed nation, and yet our national newspapers, with just a few honourable exceptions, provide little more than a bare outline of the day's forecast. I am therefore immensely grateful to a succession of Editors of the *Sunday Telegraph* for encouraging a gradual growth over the years in the size of the paper's weather column; in particular, the present incumbent, Mr Charles Moore, has given me more space and freedom than ever before to write about things other than the coming day's weather.

This book incorporates some of what I have written over the years in the *Sunday Telegraph*, plus much more.

It is aimed at anyone who just enjoys reading about the weather, and it assumes no scientific knowledge, so the occasional attempts to explain physical processes are kept very simple. Even so, the more scientifically minded reader will, I hope, find something entertaining here too. *Weatherwise* comprises a series of linked essays, so it can be enjoyed either by reading it cover to cover or by dipping into it at random.

This is not meant to be a meteorological text or reference book and the choice of topics was very much a personal one, so some people's favourite weather events may not have found a place. But the facts and figures have been carefully checked, and in *this* volume at least, all the statistical references can be regarded as reliable.

I would like to acknowledge all the many sources of original published material that I have used, in particular the *Journal of Meteorology*, *Weather*, the *Meteorological Magazine* and its predecessor, *Symons's Meteorological Magazine*, *British Rainfall*, and the *Monthly Weather Report*. And special thanks are due to Stephen Burt for double-checking the facts and figures and for his encouragement throughout the project.

1. The Weather Machine

Air, fire, earth and water were the elements according to the ancient Greek philosophers. When we learnt basic chemistry at school and became familiar with the periodic table of the elements, we were able to laugh down our sleeves at the ancient Greeks: "We know so much more than they did and we're only twelve years old." But they weren't so daft really. Substitute the sun for fire, and they did in fact identify the essential components of the weather machine.

Without air, warmth and water planet earth could not support life, but without any single one of those elements, weather as we know it simply would not exist. The air is the medium in which all weather happens, heat from the sun provides the energy that drives the weather machine, and water provides much of the variety of our day-to-day weather. Without water we would have no clouds, no rain or snow, no hail, no mist or fog, no dew or frost, and probably no thunder or lightning either. All we would be left with would be sunshine, wind, heat and cold, and the occasional duststorm.

There was a time, before the electric telegraph brought the first almost instantaneous communication, when most ordinary people were unaware that the weather events in any one place were part of a bigger pattern. It might rain in Birmingham between seven and eleven o'clock in the morning, and in London between midday and four in the afternoon, but no one knew that both downpours might have been caused by the same rain belt, or indeed that hundreds of other towns and cities across the country were experiencing similar weather at different times. Some scientists were beginning to make connections, though, and when the telegraph came into regular use, the collation of meteorological information allowed early meteorologists to make a synopsis of weather conditions at roughly the same time across a large geographical area – the prototype "synoptic" chart.

The 1880s were a very exciting time as meteorologists first discovered that some severe storms (depressions) experienced along the eastern seaboard of the USA travelled across the Atlantic, tracked by collecting masses of ships' reports, and appeared in western Europe several days later. The hope and expectation was that it would be possible to follow the travels of the highs and lows across the daily synoptic charts, enabling forecasters to predict the weather weeks ahead. The 1890s brought the inevitable disappointment, as more and more data showed that some depressions did indeed cross the Atlantic, others got so far and then died, and yet others formed in mid-Atlantic and swept across the British Isles virtually unannounced. The global weather machine was far, far more complicated than anyone had ever believed.

This sort of forecasting – using surface weather charts, empirical rules, and a fair measure of intuitiveness – was useful at predicting the weather up to 24 hours ahead and was occasionally successful for a 48-hour period. This was the norm in the 1930s. When regular upper-air information became available in the 1940s and 1950s, these empirical methods reached their peak, with improved 24-hour forecasts, useful 48-hour forecasts, and comparatively rare successful ones for up to four days ahead.

Decades ahead of his time, Lewis Fry Richardson devised numerical forecasting techniques during the early 1920s – predicting the behaviour of the atmosphere mathematically by applying the laws of physics. Unfortunately for him, it was impossible to use numerical forecasting during his day because the quantity of mathematical calculations required to produce a 24-hour forecast took far too long to complete manually. Not until computer technology began to take off in the 1960s did Richardson's ideas really begin to bear fruit, and it was not until the late 1970s that the computer-generated forecasts began regularly to outscore those produced solely by humans. Even today, the most consistently successful forecasts are produced by computers with extensive supervision and frequent intervention from the country's best human forecasters. This joint effort is called, rather unattractively, the man-machine mix. In the 1990s, a high level of accuracy is achieved up to three days ahead, and useful guidance for six or seven days. At last the dreams of the 1880s are beginning to come true.

WHERE DO CLOUDS COME FROM?

People have studied the skies from time immemorial, sometimes in sheer wonder at the beauty or drama of cloudscapes, sometimes searching for signs of changes in the weather. There is an infinite variety of cloud shapes, sizes and colours, but some formations do tend to recur, some more regularly than others. Our ancestors doubtless associated these with different types of weather, and they probably recognized some as precursors of a change in the weather – the first weather forecasts.

To a meteorologist, cloud types fall into two main categories, layer clouds and lumpy clouds – which he would call "stratiform" and "cumuliform" respectively. A non-meteorologist would also recognize two categories, but they would be different – rain clouds and fair-weather clouds. Unfortunately the two classifications do not match; some layer clouds produce rain while others don't, and the same is true of cumuliform clouds. Nevertheless, an experienced cloud-watcher, even without any other meteorological training or information, can probably tell at least 90 per cent of the time which clouds will produce rain within an hour or two and which clouds won't.

There are two simple rules of physics that help us to understand a little about how clouds form. Firstly, warm air rises, and as it does so it gradually expands and cools; secondly, the warmer the air is, the more moisture it can hold. All air contains water vapour in some measure. Even in the middle of the Sahara desert, where it might not have rained for months, on any given afternoon there is probably more water vapour held invisibly by the air than there is in the air over England. As an illustration of the degree to which warm air can hold more moisture than cold air, we can calculate that the quantity of water vapour that saturates the air at 10 °C (50 °F) is only 17 per cent of the quantity needed to cause saturation at 40 °C (104 °F), so for the same amount of moisture the respective relative humidities are 100 per cent and 17 per cent. The moisture in the desert air sometimes makes its presence felt at night when the temperature falls sharply and a heavy dew may condense on the ground.

Now, air is sometimes forced to rise, either by flowing over a range of hills, or flowing over a large bank of colder air, or when it is warmed from underneath. As the air rises it cools, and as it cools its relative humidity rises. Eventually it will reach a point where the relative humidity is 100 per cent, the parcel of air can no longer hold all its moisture, and some of it condenses into (visible) water droplets, thus forming an embryonic cloud. Depending on the temperature structure of the air above the ground, the clouds may continue to grow into rain clouds, either layered or lumpy; they may remain as fair-weather clouds; or they may dissipate again almost as soon as they form. Bubbles of air warmed from underneath, usually the result of strong sunshine heating the ground, which in turn warms the air in contact with it, will form cumulus clouds, initially like small chunks of cotton wool. If the bubble is very buoyant it will continue to rise under its own momentum, accompanied by neighbouring buoyant bubbles, and the cumulus cloud will grow bigger and bigger, eventually climbing several miles high and turning into a cumulonimbus cloud and threatening showers and thunder.

WHAT MAKES IT RAIN OR SNOW?

The water droplets in a cloud are so tiny that they stay suspended in the air, largely unaffected by gravity. In fact that is not quite true, but they may fall at only 20 or 30 feet per hour. The droplets will be of slightly different sizes and the bigger ones will fall slightly faster than the smaller ones, colliding with and absorbing some of the smaller ones. The process accelerates, and eventually the droplet will become big enough and heavy enough to create a rain drop.

There is another mechanism that works more efficiently, but it is rather more complicated. Most clouds in the latitude of the British Isles are at least partly at a temperature several degrees below freezing. Cloud droplets can exist in liquid form at these temperatures, but ice crystals will coexist, and these ice crystals attract water vapour more effectively than the water droplets do. Thus the ice crystals grow at the expense of the water droplets, eventually forming snowflakes big enough to fall out of the cloud. As the snowflakes drop below the freezing level they will melt into raindrops. This is the more usual of the two rain-making methods over the British Isles.

This also disproves the common misapprehension that snow is frozen rain. As can be seen, it is precisely the other way round.

WHERE DO THUNDER, LIGHTNING AND HAIL COME FROM?

Thunder, some of us were told at our mother's knee, is God moving his furniture about. Meteorologists wish it was that simple. Even today, all the intricacies of atmospheric electricity are not properly understood, but the experts now know quite a lot about what makes a thunderstorm.

In a large cumulonimbus cloud, the vertically moving air currents – the updraughts and downdraughts – are so powerful that large water droplets can be carried up and down within the cloud several times. The upper part of the cloud is usually cold enough for the water droplet to freeze into a small hailstone, which then grows larger on each upward journey until it is finally too heavy to be carried upwards again, or more usually it simply fails to find a sufficiently strong part of the updraught and just falls out of the cloud. It is believed that small hailstones colliding with tiny ice crystals create an electrical charge – a negative charge on the hail pellets which fall into the lower part of the cloud, and a positive charge on the ice crystals which are carried into the top of the cloud by the updraught. The Earth's surface and objects on it also carry a positive charge.

Electrical discharges therefore occur between the negative and positive charges, both within the cloud, and between the bottom the cloud and the ground. All lightning is what is commonly called "forked", but if the electrical activity occurs inside a dense cloud or is a long way away, the discharge itself may be invisible and all you may see is a reflection of it on the cloud itself or on other neighbouring clouds. This is what is called "sheet" lightning. When lightning strikes the ground, the first thing that happens is a channel of least resistance is created downwards from the cloud. This is called the "negative leader" because it carries negative charge from the cloud to the ground. As soon as it meets the positive charge on the ground, or more usually on a tree, church spire, chimney pot, or other protruding object, the "return stroke" occurs and this is the brightest part of the lightning stroke. Sometimes there is too much charge for one stroke to release, and several strokes will occur along the same channel, creating a flickering lightning flash which may last as long as a second or two.

A lightning flash carries an enormous electrical current, often between 10,000 and 50,000 amps, which produces some 30 megawatts of power per foot, and this in turn heats the air along the lightning channel to a temperature of roughly 30,000 °C. This causes a violent expansion of the air – an explosion really – which is heard as thunder. Thunder travels at the speed of sound, and that allows us to estimate how near the lightning flash occurred. A count of five seconds from the flash to the thunder means that the discharge happened roughly a mile away, but you also have to remember that the thunder cloud may be a mile or more above the ground. If the discharge was very close, the lightning and thunder will happen almost instantaneously, but after a huge initial crash the thunder will rumble on for several seconds. This is because the sound from the upper part of the lightning stroke will take longer to reach you than the sound from the lower part of the stroke.

2. The British Climate

What makes the British climate so British? Rain falls on London and Leeds just as it falls on Lisbon, Lagos and Little Rock. The sun shines out of the same sky and the wind blows in the same way. So what constitutes the British climate, and why is it so distinctive? The answer lies in the pattern of weather as it proceeds day by day through the months of the year – the way it rains little and often, the large number of dull but dry days, the frequent changes, but the rarity of extremes and dramatic switches in the weather. Someone once said that climate was nothing but a large collection of daily weather, and someone else suggested that Britain had no climate, only weather. But, in truth, the very variability of our weather from day to day throughout the year, and the differences between one year and the next, provide the essential characteristics of the British climate.

There are of course large differences in the climate across the British Isles, from the "Atlantic" climate of the mountainous western seaboard where it rains a lot, the wind blows a lot, and the sun shines reluctantly, to the relatively dry regions of south-east England and East Anglia where the winters are colder and summers warmer than most other parts of the country. But these differences are quite small in comparison with the climatic contrasts that can be found across relatively small countries like France and Spain, let alone larger ones like the USA or Australia, or between one end of Europe and the other.

The "British" type of climate is not quite confined to the British Isles. Apart from adjacent parts of Europe like northern France, Benelux and Denmark, similar climates are found around Seattle and Vancouver, in parts of New Zealand and Tasmania, and in parts of southern Chile. All these areas have one thing in common – their geographical siting. They are all situated in the temperate-latitude belts of westerly winds with large tracts of ocean upwind. But nowhere quite matches the geography of northwest Europe, where there is no major mountain barrier to hinder the westerly winds as they blow onwards into the continent, and where the North Sea plays a very important role in ameliorating any continental weather when the westerly winds occasionally give way to easterlies. If the North Sea were not there, the climate of eastern England would be colder in winter, somewhat hotter in summer, and drier throughout the year. Eastern Scotland would be much colder in winter, much warmer in spring and summer (especially spring), and very much drier – indeed Aberdeen would probably be one of the driest cities in Europe.

Another interesting if fruitless exercise is to imagine what sort of climate Britain would have if the earth rotated in the opposite direction. Then the dominant wind belts of our latitudes would be easterlies, and we would find ourselves in a position presently occupied by Sakhalin Island or Kamchatka in the Russian far east, or by Newfoundland in North America. The winters would be so cold that the North Sea would probably be frozen for four or five months of the year, spring would come late and it would be damp and sometimes foggy, and summer would be short and rather cloudy and misty. Sheltered parts of western Britain – the west coast of Scotland for instance, Morecambe Bay, or the Cardigan Bay coastline – would probably have the nicest summer weather, but these same areas might suffer the heaviest of the winter snowfalls. All in all not very enticing, is it?

Britain lies at a sort of meteorological crossroads. Although the main traffic arteries approach from the west, winds from a westerly quarter blow for only 30 to 35 per cent of the time. Southerlies are almost as common, averaging 25 to 30 per cent; easterlies blow for 15 to 20 per cent of the year; and northerlies for about 20 per cent. These winds are not just local breezes; they represent mass movement of air originating thousands of miles away towards the British Isles, and several characteristic air masses can be identified according to their source region. These contrasting air masses do not mix very well when they come into contact with each other, and the boundaries between them are usually zones of bad weather, which are marked on the TV weather maps by "warm fronts" and "cold fronts". You could liken these fronts to traffic lights. For instance, several days of southwesterly winds may bring a long stream of tropical maritime traffic across Britain from the region of the Azores, say, or Bermuda; a cold front then crosses the country, cutting off the southwesterly flow, the traffic lights change, and a stream of polar maritime traffic races in from the northwest behind the front.

These huge travelling masses of air are given names that describe their origins: either "maritime" or "continental", depending on whether the source region is over the ocean or over the land, combined with one of "tropical", "polar", or "arctic", which describes roughly the latitude of the air mass's origin.

Polar-maritime air is the most frequent visitor to the British Isles. It originates over the extreme north of the Atlantic Ocean in the vicinity of Iceland or Greenland or northeast Canada, and it usually arrives on a northwesterly or westerly wind, having travelled across relatively cool parts of the ocean. It usually brings a mixture of sunshine and showers, with cumulus-type clouds and blustery breezes. Haze and pollution levels are very low, so everything looks freshly scrubbed: the colours of the countryside show up clearly, visibility is excellent, and the horizon stands out sharply. Between late autumn and early spring most of the showers occur in the west and north, but during the summer half of the year the showers occur widely across inland areas while coastal fringes often escape. Temperatures at all seasons are near or rather below average, with night frosts during the winter and spring, but unusually low temperatures are very rare.

A variant of polar-maritime air occurs when the air arrives via a rather circuitous route, initially plunging southwards into mid-Atlantic then curving eastwards and then northeastwards, approaching the UK from the southwest. In other words, the polar-maritime traffic is diverted onto the road that usually brings tropical-maritime traffic, but it still retains its "polar" characteristics although it may have been modified en route; this is called "polar-maritime returning". This is still a showery air mass, although the air may be warmer and moister – and therefore rather cloudier and hazier – than true polar-maritime. Temperatures are usually near or slightly above the seasonal average at all times of the year.

Southwesterly winds normally bring a tropical-maritime air mass, which has travelled to us from sub-tropical latitudes of the Atlantic Ocean. It therefore starts off very warm and very moist, but the whole of its journey to the British Isles is across progressively cooler parts of the ocean, so the lower layers of the air are gradually cooled. This sort of set-up, with relatively cool air at low levels and warmer air aloft, is highly stable in the sense that upward and downward moving air currents are effectively damped down, so the airflow is very stratified. This means that clouds, too, are usually stratiform (layer clouds) and the sky is often grey and featureless while the weather may be damp and misty and drizzly, especially around the coasts and hills of the south and west that the airflow first encounters on reaching the British Isles. If the air is particularly warm and moist through a considerable depth of the atmosphere, the uplands of western Britain may get prolonged steady rain, entirely induced by the forcing of this air mass over the mountains. In summer, the air may be dried sufficiently as it crosses the hills of western Britain for the layer of stratus cloud to break up, and on these occasions it becomes quite hot in the Midlands and eastern Britain. Broadly speaking, though, tropical-maritime air is best described as mild and muggy.

True tropical-continental air originates over North Africa, or perhaps over Spain. In practice, it is virtually impossible to differentiate between North African air, which has usually crossed the western Mediterranean, and air that has originated over the Mediterranean Sea itself. This air mass is often humid and thundery and would perhaps be better described as "Mediterranean" rather than "tropical continental". It is always very warm between late March and mid-October, sometimes exceptionally hot, and is responsible for most of the high-temperature records established during the summer half of the year. It is a rare visitor during the late autumn and winter, simply because winter weather patterns do not usually favour the development of long streams of southerly or southeasterly winds across western Europe. When it does appear in winter, tropical continental air is usually much modified by its passage across France, which is part of the the cold continental land mass at this time of the year. It may be mild and dry but probably also grey and very misty with high pollution levels.

"Polar continental" is the name given to air that originates over Europe and that reaches us from the easterly quarter. The precise wind direction is crucial, because this determines how long the air mass lingers over the North Sea on its journey from the European mainland, and therefore how much modification takes place as it crosses the sea. Clearly, air arriving on a southeasterly wind is modified least, easterly air is modified a lot in northern Britain but comparatively little in the south, and northeasterly air is substantially modified in all regions. This air mass is not really "polar", and would perhaps better be described as "European". During the winter half of the year it is a very cold, dry air mass, sometimes with cloud-free skies and sometimes with a shallow layer of stratus cloud, while during the summer half-year it is quite a hot air mass, usually with clear skies. At all seasons it is hazy with high pollution levels. The North Sea plays a crucial part during the winter because it lifts the temperature of the air by several degrees, although at the same time it may inject sufficient moisture in the lower layers of the atmosphere to produce snow showers. In spring and summer, the North Sea provides a cooling influence, and the added moisture may result in persistently dull and cold conditions in eastern Britain, while sheltered western districts enjoy unbroken warm sunshine. This effect probably reaches its peak during May when the temperature difference between the cold North Sea and the warm European air mass reaches its maximum. On such occasions places like Aberdeen or Tyneside may struggle to climb above 6 °C (43 °F) under leaden skies and with a raw wind off the sea, while on the lee side of the mountains in Wester Ross and Lochaber and in Cumbria and north Lancashire brilliant sunshine and a warm breeze can send the temperature rocketing to 25 °C (77 °F) or higher. Regrettably for east-coast counties, this polar-continental or European air mass occurs most frequently during late winter and spring.

Arctic-maritime and arctic-continental air, as their names imply, originate within the Arctic Circle, the former usually coming from the north Norwegian Sea between Spitzbergen and Greenland, and the latter travelling to us from northern Scandinavia or northern Russia. Arctic-continental air is responsible for the weather that is dubbed "Siberian", although air originating on the far side of the Ural mountains very rarely if ever reaches the British Isles. Both these air masses occur mostly during winter and spring, bringing our coldest weather, frequently accompanied by snow. Arctic maritime is really just a colder version of polar maritime, and the two are barely distinguishable in summer. In winter, though, the direct route from the Arctic with a relatively short track across the warm sea means that wintry showers are often confined to north-facing coasts and hills, while other parts of the country stay sunny but bitterly cold. Arctic-continental air is fundamentally very dry, but some moisture may be picked up over the Baltic Sea as well as the North Sea, and substantial snowfall can occur in eastern counties of both England and Scotland.

Sometimes, Britain develops its own air mass. When

a high-pressure system lingers for several days over the British Isles, the air becomes stagnant and takes on the characteristics of the season. Thus "home-grown" air will be cold and dry in winter with night frosts, frequently accompanied by mist or fog or low cloud, while between roughly late March and mid-October it will be dry and reasonably warm.

3. January

"Generals January and February will fight for us."
TSAR NICHOLAS I, 1853

Those two Russian generals are certainly very formidable characters. They broke the back of many a military campaign, and can probably be credited with ending the expansionist aspirations of the likes of Napoleon and Hitler. But General January's British cousin is of a much lower rank – a rather bolshie corporal, perhaps – an infernal nuisance at times but rarely any serious trouble.

Most Januarys in the British Isles bring a mix of contrasting weather types, thanks to our position in temperate latitudes on the Atlantic margin of the great Eurasian land mass. These weather types can largely be described in terms of their source regions, and therefore are usually dubbed tropical, polar, arctic, continental, maritime, or some combination. In mid-winter, the contrasts between ocean and continent and between the tropics and poles are at their greatest, hence the abrupt changes that can occur in the weather at this time of the year.

The character of an entire month depends upon which of the various weather types are dominant. We learn from our school geography books that the prevailing wind direction in our part of the world in winter is southwesterly (that is, *from* the southwest). Sure enough, the records show that the dominant wind in three out of four Januarys is a southwesterly one. It won't blow from that direction on every day during the month, of course, but it does mean that a substantial proportion of January days have a wind blowing from warm-temperate or sub-tropical latitudes over the Atlantic Ocean, and that is why so much of our weather at this time of the year is moist and mild in comparison with similar latitudes in Europe or North America. Occasional Januarys are characterised by airstreams from the north, the east, or the south, or they may be dominated by anticyclones (high-pressure systems) or depressions (low-pressure systems) with no dominant wind direction. The one really outstanding example of a "non-westerly" January occurred in 1963. During that month there were 20 easterly days, while the remainder were northerly or anticyclonic, with not a solitary westerly day in sight. As a consequence it was the coldest January in the last 250 years, and much of England and Wales was snow-covered throughout.

January has produced its fair share of weather disasters during the last few centuries. Floods and gales figure most prominently in the chronicles: the infamous North Sea flood on 31 January and 1 February 1953, which took over 2000 lives in Britain and the Netherlands, accompanied by a destructive northerly gale over Scotland, was our biggest natural disaster of the century; the "night of the big wind" on 6/7 January 1839 is considered to have been the most destructive storm ever to have hit Ireland and the death toll on both sides of the Irish Sea probably exceeded 400; a naval yacht foundered in a gale off the Scottish coast on 1 January 1919 with the loss of 200 lives; the Glasgow hurricane of 15 January 1968 killed 20 people and caused damage to a third of a million properties; further destructive gales resulting in substantial loss of life swept the country on 2/3 January 1976, 11/12 January 1978, 25 January 1990 (the Burns' Day Storm), 5 January 1991, and at intervals during January 1993. In other months, severe wintry weather paralysed transport, seriously disrupted the nation's economic activity, and caused an increase in the death rate from hypothermia. Alongside the extreme month of January 1963 we may note in particular the blizzard of 18–21 January 1881, and the combined snowstorm and ice-storm of 26–30 January 1940.

THE GREAT SNOWSTORM OF JANUARY 1881

Imagine you are travelling down to London from Manchester by train. It is a ferociously cold January afternoon as you head for London Road Station (now Piccadilly), the east wind whistling through the near-deserted streets, and the thin watery sunshine of the morning has given way to slate-grey skies although it is still dry. The train accelerates through the Manchester suburbs and out into the wintry Cheshire countryside, still patchily snow-covered from a snowfall about a week earlier. As the train calls in at Crewe Junction, you think you notice the first tiny flakes in the fading light, and the platform at Stafford is alive with swirling spirals of dry powdery snow – quite eerie in the guttering gaslight. You shiver momentarily, and, looking forward to reaching your elegant terraced home in Camden Town by eight or nine in the evening, you snooze for an hour or so in the warm cocoon of the compartment.

You awake with a start. The train has stopped; the compartment is cold. Through the window you can see nothing but white. A dizzying maelstrom of snowflakes is accompanied by a moaning wind. In fits and starts the train inches forward, taking another hour to reach the comparative civilization of the station at Leighton Buzzard. There is some talk amongst your fellow passengers of leaving the train and seeking accommodation in the town, but the station is deserted, the up platform is already piled high with drifts, and the consensus is that the train should reach London by late evening. The discussion ends as the train lurches forward again.

South of Leighton Buzzard the main LNWR line enters a very deep cutting, four miles long, in the run up to Tring station.

Earlier trains had sneaked through the growing drifts, but yours is the first to fail. After the most uncomfortable night of your life, fearful of ending your days in the icy wastes of Tring cutting, and hunger gnawing, progress resumes fitfully around nine in the morning. Thanks to the heroic unseen efforts of the railway company, some rudimentary snowploughs, and huge gangs of company labourers, your train finally limps into Euston at half past ten, precisely 14 hours and 30 minutes late.

The year is 1881. The date is now Wednesday 19 January. It is still snowing and the wild east wind is still raging. Outside Euston Station, London is practically unrecognizable – the virtually empty streets almost silent. It is not so much the general depth of snow – probably no more than six inches, but some places are swept completely clear whereas elsewhere massive drifts are piled four or more feet high. Visibility is intermittently reduced to a few feet as the violent east wind carries away the tops of the drifts in dense swirling clouds of snow. The mile-long journey on foot back to your house in Camden Square takes another hour, and, to add insult to injury, you discover the dry powdery snow has penetrated through the cracks around doors and windows, a small snowdrift sits mockingly in the hallway, and meltwater trickles from the window-sills onto the morning-room floor.

The snowstorm of 18/19 January 1881 was arguably the worst of the 19th century. These days it would undoubtedly be called a blizzard in view of the powdery nature of the snow and the severe gale that accompanied it, but the word was not is common usage in the UK in 1881. There had been heavier falls of snow in 1814 and 1836, while the Great West Country Blizzard of March 1891 was at least as bad in Devon and Cornwall. But for the combination of deep snow and high wind, this storm was in a class of its own. No snow fell north of a line from Liverpool to Scarborough, but south of a line roughly from Bristol to London and also over the Cotswolds the fall amounted to more than 10 inches. Worst-hit counties were Devon, Dorset, Wiltshire,

Hampshire and the Isle of Wight, with level snow (so far as it could be measured) approaching two feet deep. Giant drifts covered the sides of houses, completely buried hedgerows and filled the lanes between them.

The Isle of Wight was dealt a particularly bad hand, as the island was swept by a second snowstorm two days after the first, which doubled the depth of snow. At Newport the first fall was measured at 16 inches, and the second at 18 inches. Drifts in the neighbourhood of Cowes, and also in the west of the Isle were said to be 12 to 15 feet high. At Ryde the gas lamps in the main street were left alight as it was impossible to reach them, at Shorwell the village school was completely buried and a tunnel had to be dug from the front door to the road to allow the schoolmaster and his family to be rescued, and at Chale there was no bread or flour delivery for eight days. The Isle of Wight railway was not fully operational for nine days, and many country roads remained blocked for a similar period. South Hampshire was affected almost as harshly, with Portsmouth completely cut off for a total of 38 hours, while some roads through the Downs were impassable for a fortnight.

The wind itself was responsible for much misery: several ships off the east coast and in the Channel were sunk, thousands of roofs were damaged, hundreds of thousands of trees were uprooted, and telegraph wires were brought down in innumerable places. The gale also caused a tidal surge up the Thames Estuary, inundating thousands of homes in London, chiefly along the south bank in Bermondsey and Southwark. Both before and after the snowstorms came intense cold with night-time temperatures dropping below −20 °C (−4 °F) on several occasions in southern Scotland and northern England. Lowest of all was −26.7 °C (−16 °F) at Kelso on the night of the 16th/17th. But a thaw finally set in around the 26th/27th, although remnants of drifts remained for many weeks thereafter.

THE WINTER OF '62–'63 . . . YOU REMEMBER . . . THAT WINTER!

A lot of people, of course, don't remember the infamous winter of 1962–63 – they are not old enough. They may have heard about it, but there is an abundance of myths, exaggerated stories and faulty memories, so it may be difficult to get a balanced picture of what that winter was really like.

It is worth briefly reminding ourselves what life was like at the end of 1962. Harold Macmillan's "never had it so good" era was drawing to a close, the world had held its breath during the Cuban Missile Crisis in October, and the Berlin Wall was just one year old. The Beatles were on the point of breaking through to national recognition, while we laughed with TV favourites like Tony Hancock and *Steptoe and Son*. Most people didn't own a car or a telephone, and very few homes enjoyed the benefit of central heating. In fact most people travelled to work by bicycle or public transport, or on foot, while most people's houses were warmed by coal fires, but smokeless zones had been designated in many of our major cities.

According to the statistics, this was the coldest winter of the century over practically the whole of the British Isles and a large part of the near-continent. Over central England, where comparable figures extend back over 330 years, it was second only to the winter of 1739–40, although there was less than a tenth of a degree between them. There had been a week of snowy weather in mid-November and a spell of severe frost and smog in early December, but the true wintry weather set in around 22 December and continued without a break until the beginning of March. It wasn't the snowiest winter on record in terms of quantity, but the ground remained snow-covered over a large part of the kingdom from 26 December until 2 March. And this was easily the longest such unbroken period ever recorded. So, yes, the winter of '62-'63 was a record breaker.

England and Wales were sunny but bitterly cold, with a

biting northeast wind for several days up to and including Christmas Day. Early-morning temperature on the 25th fell to −12.2 °C (10 °F) at Warsop in Nottinghamshire, while RAF Shawbury in Shropshire recorded a maximum temperature on Christmas afternoon of −3 °C (27 °F). At St Mawgan, near Newquay, on the normally temperate Cornish coast, the temperature remained below freezing all day in spite of 7 hours and 25 minutes of bright sunshine. Scotland, meanwhile, was milder and cloudier with outbreaks of rain, but the rain turned to snow for a time, especially on Clydeside and over the Southern Uplands, and Glasgow had a white Christmas. The snow spread southwards to reach Manchester and Leeds by midnight, and Birmingham and Nottingham by dawn on Boxing Day. In London and the southern Home Counties it began snowing around mid-afternoon on the 26th and continued with hardly a break until the early hours of the 28th, leaving a cover 12 to 18 inches deep over much of Kent, Surrey, Sussex and Hampshire. After a brief respite, snow returned late on the 29th, this time accompanied by an easterly gale, and the following 24 hours saw the worst snowstorm of the whole winter as heavy drifting snow swept across the whole of England, Wales and southern Scotland. Innumerable towns and villages were cut off by drifts 20 feet or more high, road and rail transport was severely disrupted, and the thaw that was predicted to follow the snow only reached Cornwall and Devon.

Further snowstorms occurred over large areas on 3/4 and 19/20 January, and a particularly severe one battered southwest England, Wales and Northern Ireland on 6/7 February. On the latter date, level snow was reported to be five and a half feet deep at Tredegar, high in the Welsh Valleys in the northwestern corner of Gwent. But between these well-scattered snowfalls, the weather was predominantly bright with plenty of sunshine and hard frosts, and occasional days of freezing fog. Great fears were voiced as to how the winter would end. The very snowy winter of 1947 ended with the worst inland flooding of the century, accompanied by severe gales. But the winter of 1962–63 gently relinquished its grip during the closing days of February and the first few days of March. Over a period of more than a week, brilliantly sunny days alternated with clear frosty nights, but a gentle thaw in the sunshine allowed the snow to disappear almost imperceptibly.

The extraordinary persistence of easterly winds resulted in some areas escaping much of the wintry weather. St Mawgan in Cornwall, sheltered by both Bodmin Moor and Dartmoor, recorded 114.4 hours of bright sunshine during January, easily the biggest total ever recorded for the month in the UK. Similarly, February's sunniest spot was Sellafield, in the lee of the Lake District, with 135 hours. Ambleside, in normally snowy Lakeland, had only three days with snow cover in January, while Prestwick Airport on the Ayrshire coast had only six such days in the entire winter. Fort William, usually one of the wettest towns in Britain, had only 3 per cent of its normal rainfall in January, and none at all in February. Brush fires were reported from the Scottish Highlands, and Cape Wrath – the northwesternmost tip of Scotland – logged a relative humidity of just 6 per cent on 2 March.

IF ONLY HITLER HAD KNOWN

During the Second World War weather forecasts vanished from the radio and newspapers for the duration. Talking about the day's weather on the air was not permitted, and reports of unusual or extreme weather in the papers only appeared six weeks after the event. Employees of the Met Office were transferred from the civil service to the RAF, and their work was shrouded in official secrecy. Weather forecasters plotted their charts on tracing paper with no geographical identification, so that the charts would be meaningless in the unlikely event of their falling into the wrong hands. In order to give weather briefings these tracings were superimposed on base-maps showing coastlines and airfields and so on.

From the comfort of the 1990s, this might seem like overkill, but in fact it was very necessary. One example early in the war provides an astonishingly vivid illustration of why this was so. January 1940 was, overall, the coldest for over 100 years – since 1838 to be precise – although it was subsequently overhauled by that of 1963. Between 26 and 29 January Britain was ravaged by a combined snowstorm and ice-storm that left the greater part of the country paralysed and from a military point of view extremely vulnerable.

The snowstorm itself was the worst since the Great Christmas Blizzard of 1927, but what was really exceptional about the last week of January 1940 was the ice-storm – that is, a widespread, heavy and prolonged fall of rain with the air temperature well below freezing, coating all objects in clear ice. This phenomenon is rarely severe in Britain, and on most occasions quickly followed by a substantial rise in temperature, so the hazard soon passes. It happens more frequently in the United States, particularly in New England and the Mid-Atlantic states, where millions of dollars of damage can occur quite regularly as a result. Its occurrence is also rather more common in continental Europe than it is in Britain. Freezing rain may occur towards the end of a cold snap, when advancing warm air rides above a stagnant mass of cold air. If the warm air is sufficiently warm, rain rather than snow may fall, penetrating the shallow cold layer and remaining liquid until impact. The coincidence of circumstances required for freezing rain means that only a narrow zone is affected at any one time, and if the weather systems are moving at any speed, the freezing rain will soon give way either to snow, or to a thaw. Only if the weather system responsible is practically stationary will the build-up of ice be significant. This sort of ice is known in the weather business as glaze; other terms include glazed frost, and, less correctly, silver thaw, and black ice.

The worst effects of the January 1940 ice-storm were found in a broad swathe from Dorset, Hampshire and west Sussex in the south to the west Midlands and mid-Wales in the north.

Freezing rain also fell as far east as Kent, Essex and Cambridgeshire, as far west as Devon and the west coast of Wales, and as far north as Cheshire, although in these peripheral areas damage was much less serious.

This exceptional storm was initially investigated in some detail by C.J.P. Cave, who published an account shortly afterwards, including eyewitness descriptions from a variety of correspondents. Cave himself lived at Stoner Hill, near Petersfield, on the Hampshire–Sussex border. He wrote:

> The formation of ice began at 4 pm on the 27th. By next morning everything was covered with ice, the depth on the thermometer screen being nearly half an inch. It was almost impossible to walk; paths and roads were covered with a smooth coating; on the grass the ice was more irregular but it was almost as difficult to walk on. The vicar of a neighbouring parish told me that he went to church on his hands and knees. It went on raining more or less all day on the 28th, and many windows became so stuck by the ice that they could not be opened. In the afternoon branches began to fall from some tall poplars, and telephones were out of action by nightfall. The following night, branches could be heard falling, and those living closer than I do to the beech hangers said that the crash of falling boughs could be heard all night. Telephone wires 1.5 mm in diameter were encased with ice 30 mm in diameter, and twigs were covered with ice in the same proportion.

From the Cotswolds, a certain Mr W.I. Croome of Bagendon, near Cirencester, described events as follows:

> The first time I noticed anything unusual was about 10.30 pm [on the 27th]. I heard in the dark – there was little wind – a noise like castanets snapping all about me in the dark, and by my torch I discovered that every leaf on the laurels, every bunch of dried laburnum seeds and sycamore seeds was encased in ice and rattling in this uncanny way. At 9.45 am on Sunday the 28th we first heard the crash of falling timber from the wood opposite, followed by

> another. By lunchtime the crashes were occurring every few seconds, and by nightfall this valley resounded with a constant, horrible roar of falling timber, as if we were under bombardment. It kept us awake all night. A small spray of beech twigs, ice-loaded, was weighed and proved to be 3.5 lbs; then the ice was melted and the twigs again weighed – 3.5 ounces! That gives 15 times their weight in ice. An investigation of telephone wires . . . gave a result of just over 11 tons of ice between two poles. No wonder they snapped like matches. Nothing looked lovelier than the common elder bush. It remained upright in its gleaming transparent casing, like huge fantastic crystal candelabras. The beeches looked like frozen waterfalls; the long sprays were fastened together with ice, then hung downwards by the weight. Here the ice lasted quite intact until February 3rd when at last it melted. On Sunday the 4th, the scenes of flooding were incredible. Every inch of ground was solid ice, and water poured down the frozen slopes into the valleys. Here, the whole valley floor was a raging torrent 4 to 5 feet deep, over the tops of garden walls, and the church was flooded to the tops of the pews.

In recent years, occurrences of glaze have been practically non-existent, but freezing drizzle caused a little local difficulty in the Midlands and southern England in January 1987, while the hills of the Midlands and northern England were quite badly affected in early February 1986. It is practically impossible to gauge how exceptional the 1940 ice-storm was, but it was so much more extreme than anything else in the United Kingdom in the 20th century that we may not have anything quite as bad for another 100 or 200 years.

TO SNOW OR NOT TO SNOW

Weather forecasters have a love-hate relationship with snow. Many of us have a little-boy attitude to the stuff, longing every winter for a really good fall, in spite of all the inconvenience and expense that can ensue. On the other side of the coin, forecasting snow can be inordinately difficult, especially when the temperature is near or above freezing, and in Britain the majority of snowfalls occur on such occasions.

On near-freezing days, the choices open to the forecaster are legion. Having first decided that precipitation is likely, he next has to choose between rain, drizzle, sleet, snow, and freezing rain/drizzle. There are also some rather exotic forms of precipitation which turn up on odd occasions, such as ice pellets, soft hail, snow grains, ice needles, and of course "ordinary" hailstones of varying dimensions, but these will not usually enter the equation at this stage. Then he has to consider whether the temperature is changing, and therefore whether the snow is going to turn to rain, or vice versa. But the precipitation itself may change the temperature of the air that it is falling through, and that in turn depends on four things – the temperature of the air between the cloud and the ground, its humidity, the strength of the wind, and the heaviness of the precipitation.

Nor is the decision between different forms of precipitation always confined to occasions when the temperature is near freezing point. It has been calculated that under exceptional atmospheric conditions a snowflake could just survive to reach the ground when the air temperature (as measured four feet above the ground) is as high as 8°C (46°F). An excellent example of this occurred in April 1977 when weather observers in different parts of England noted snow showers beginning with the temperature between 5 and 8 °C (41–46 °F), although once the showers got properly under way the temperature dropped sharply. By contrast, drizzle was observed falling in the Midlands on 19 January 1987 while the temperature stood at −4 °C (25 °F).

A recent example of an unexpectedly heavy snowfall which provided a splendid illustration of this forecaster's nightmare occurred on Thursday 6 January 1994. Wintry weather with a strong threat of snow had been predicted as much as five or six days in advance, but even 24 hours ahead, the detail of the snowfall was not forecast very well. Wednesday evening's forecasts for the following day indicated showers of rain, sleet or snow, especially across southern England, where some of these showers would be heavy and prolonged. This is about as accurate as it is possible to get in such meteorological situations, which are arguably the most intractable the British forecaster has to deal with. When there are sporadic areas of precipitation, constantly developing or decaying, moving erratically under the influence of light variable winds, and with ground-level temperatures a few degrees above freezing, it may be unwise and unhelpful to attempt too much detail in a forecast for 24 hours ahead. That is why rather broad-brush predictions are made on these occasions.

Throughout London and the Home Counties, the precipitation that Thursday evening started as rain, with an air temperature between 3 and 5 °C (37–41 °F), and heavy rain fell for some time before it began to turn through sleet to heavy wet snow. This had the important effect of washing away any salt and grit that remained on main roads from earlier treatments, rendering ineffective the best efforts of local authorities to keep traffic moving. Thus snow and slush accumulated readily on major routes such as the M1, the M25, and the M40 later in the evening.

So why did the rain turn to snow when the temperature was substantially above zero? Crucial were the great intensity of the precipitation and the lack of wind. According to physical laws, the temperature in saturated air decreases by 1.5 °C (2.7 °F) per 1000 feet above the ground. As the main precipitation-producing cloud mass was some 5000 feet above London, snow was leaving the cloud with an ambient temperature of about −3 °C (27 °F). The freezing level would have been at about 3000 feet, and below this point the snowflakes would begin to melt. However, large amounts of energy are required to melt ice and snow – see how long it takes to melt a saucepan full of snow on the kitchen stove – and 3000 feet above London on a chilly Thursday evening in January, the only available energy source is the air itself. This energy required to melt ice (and which is also released when water freezes) is properly known as the latent heat of fusion.

Now, a large snowflake will take some time to melt, especially as the air surrounding it will be only slightly above the freezing point, and it may fall between 250 and 500 feet as it does so. Thus, in our example, melting snowflakes between 2500 and 3000 feet above London will absorb energy from the air resulting in a progressive cooling of that layer, with the temperature eventually reaching zero, and this in turn will allow subsequent snowflakes to penetrate gradually closer and closer to the ground. The heavier the precipitation, the more energy is needed to melt the snowflakes, and therefore the more rapidly the melting level descends. The lack of wind means that there is little turbulence to disturb the development of this new temperature profile, and if the heavy precipitation lasts long enough the temperature at ground level will eventually drop to freezing, and snow will then accumulate rapidly.

From this explanation it could be expected that, assuming the precipitation fell equally heavily throughout the area, snow would begin to settle over the North Downs and the Chiltern Hills long before it did so along the banks of the Thames. But it was not that simple, because there were two or three swathes of heavier precipitation, notably one stretching from east Luton and Hitchin northwards across Biggleswade to Huntingdon and Chatteris. In this swathe, snow penetrated to the ground earlier and lasted longer than it did on either side. This meant that snow was deeper – up to six inches – over the relatively low-lying Hitchin–Biggleswade area than it was over the high ground of Dunstable Downs and Ivinghoe Beacon.

It is on such days that forecasters mutter to each other, in the privacy of the forecasting office, "What a mug's game!"

WARM, WINDY AND WEIRD

During January, snowstorms and arctic temperatures bring more disruption, inconvenience and misery than any other kind of weather. But floods and gales are not far behind, and persistent freezing fog with high pollution levels can also be a serious hazard. However, just as interesting to meteorologists are those bizarre mid-winter days that combine wind and warmth to bring unseasonably high temperatures to the most unlikely parts of the country.

There have been just seven January days in the last hundred years when the temperature has reached or exceeded 17 °C (63 °F) anywhere in the UK. The most usual national hot spots – London, Jersey, Kent, Norfolk, Southampton and Torbay – do not figure at all in the list. Astonishingly, all seven occasions occurred along the coast of north Wales, and on five of those seven days the highest temperature was recorded at the same place – the University of North Wales farm site near the unpretentious little village of Aber, located on the A55 trunk road between Bangor and Llandudno. Twice this century, a January temperature of 18.3 °C (65.0 °F) has been logged there – on the 27th in 1958, and on the 10th in 1971.

So what is so special about this tiny corner of Wales? Could it be dodgy thermometers? Or perhaps an imaginative observer? The first time such an unusual reading was reported, these thoughts may have flitted across the mind of the cynical old hand back at Met Office HQ charged with quality-controlling the figures that come from distant parts of the kingdom. When it was realized that temperatures above 14 °C (57 °F) appear fairly regularly in January, not just from Aber, but from nearby north Welsh coastal resorts such as Llandudno and Colwyn Bay, it was evident that this was a real phenomenon, requiring a scientific explanation.

What is special about the coast of north Wales is its geography. Here we have a narrow coastal plain on the north side of a substantial upland mass. The warmest winds in winter normally blow from between south and southwest, and are typically very moist. They are forced upwards over the Welsh hills, depositing copious quantities of rain, then descend the northern flank of the Cambrian mountains, perhaps drawing down much drier air from aloft. Descending air is warmed by compression, and this warming is enhanced if the air is dry. This process is generally known as a "föhn" effect – after the föhn wind, which produces abnormally high temperatures and low humidities in the Alpine region (see also Chapter Fourteen). Aber occupies a particularly favoured location for the development of a föhn, lying some 13 miles north-northeast of Snowdon – at 3560 feet the highest point in England and Wales. It is not only Aber, either. It just so happens that the University of North Wales has had a weather-recording station there for the last seventy years. It is highly likely that the adjacent resorts of Llanfairfechan and Penmaenmawr benefit to the same extent as Aber, but no official temperature records exist for those two towns.

The north Welsh coast is not unique in its geographical characteristics in Britain. Similar stretches of coastline on the north or northeast flanks of upland areas also enjoy rare days of abnormal mid-winter warmth. These include the Devon/Somerset coast to the north of Exmoor, the North Yorkshire coast around Whitby, the coastline of north Northumberland and Berwickshire, and the southern shore of the Moray Firth. As recently as 27 January 1989 a temperature of 15.5 °C (59.9 °F) was recorded at Torrisdale, roughly midway along the northernmost stretch of coastline in the country, between Cape Wrath and Thurso. Parts of the Antrim coast are also occasionally favoured.

Probably the most exceptional of all January warm spells was that of 8–11 January 1971. North Wales topped the temperature list with 18.3 °C (65.0 °F) at Aber, 18.2 C (64.8 °F) at Llandudno, 17.7 °C (63.9 °F) at Rhyl, 17.6 °C (63.7 °F) at Prestatyn, and 17.2 °C (63.0 °F) at Colwyn Bay. A few non-coastal spots in north Wales and northern England did well too, including 16.7 °C (62.0 °F) at Ceinws in Montgomeryshire, 15.9 °C (60.6 °F) at Bromfield, near Aspatria, in Cumbria, 16.6 °C (61.9 °F) at Huddersfield, 15.6 °C (60.1 °F) at Ilkley, and 16.2 °C (61.2 °F) at Corbridge in Northumberland. In the north of Scotland, Poolewe in Wester Ross recorded 16.6 °C (61.9 °F) while Lairg in Sutherland touched 16.7 °C (62.0 °F) – the highest January temperature ever recorded north of the border. Meanwhile, central London could do no better than 12.8 °C (55.0 °F) during this spell.

In one of those absurd contrasts which the British weather occasionally throws up, the temperature on the north Welsh coast five months later – on 9 June 1971 – climbed no higher than 11 °C (52 °F). So if you fancy sunbathing in Llandudno, forget flaming June, try January instead. But you will have to choose your dates carefully.

4. February

"February, fill dyke,
Be it black or be it white."

ANON

There are a whole host of different versions of the "February, fill dyke" saying, and there are several interpretations of it as well. There is no doubt that "dyke" refers to the drainage ditches found in many parts of eastern England, particularly in the Fen district of Cambridgeshire, Norfolk, and Lincolnshire, and a glance at the relevant Ordnance Survey maps shows that the area is swarming with "dykes". Many country folk these days seem to think that the old adage is merely a statement of fact, indicating that during February these drainage ditches will usually be full of either water or snow.

This, however, was not the view of Richard Inwards, one of the more eccentric of Victorian meteorologists, who was the first person to make a systematic study of Britain's ancient weather lore. He considered that it was an exhortation or even a prayer – demanding or entreating that the good month of February should provide enough precipitation to keep the dykes topped up, whether rain or snow it did not matter. The rationale is that eastern England is the driest part of the country, and spring is frequently the driest time of the year; indeed, the lack of rainfall is sometimes exacerbated by persistent drying northeasterly winds, which tend to blow for long periods between February and May. Extensive irrigation is the norm during nine summers out of ten in these eastern counties, so a thoroughly good downpour during February would help to counter the drought that almost inevitably follows, or at least to delay its onset.

These are, after all, just the things that ordinary country people in pre-Industrial Revolution times would be worrying about during late January and February, when their thoughts would be turning to the exigencies of the next year's subsistence. It is difficult to see what purpose a simple statement of fact would serve in those days. Subsequent experts on weather lore, such as Eric Hawke and Michael Hunt, endorsed Inwards' view.

February provides a surfeit of other old weather adages – another illustration of how important this month's weather was to our forebears. In fact, there are more sayings associated with Candlemas Day – 2 February – than with any other day during the year. Candlemas in the Christian Church marks the Feast of the Purification of the Virgin Mary, and it is the day when church candles are blessed, hence the name. More to the point, in the old country calendar it represented the end of the Christmas period – which centuries ago was a much more protracted if rather more subdued celebration than it is now – and the beginning of spring, no matter what the actual weather was like on the day itself. It was therefore a key date – perhaps *the* key date – in the whole year.

Probably the best known of the old saws concerning Candlemas – certainly the most often quoted – is one of Scottish origin:

> If Candlemas be fair and clear,
> There'll be two winters in the year.

This is not really meant as a prognostic; it is highly unlikely that there is a link between sunshine on 2 February and frost and snow during the remainder of the winter. It is perhaps best regarded as a general warning that winter is not over yet, even though we are now six weeks distant from the solstice and the sun is getting perceptibly stronger and higher in the sky as each new week passes.

After all, the statistics show that, over the country as a whole, February is the coldest month of the year almost as often as January is, and in some coastal areas in the west and southwest of the UK it is more likely than January to be the coldest. The temperature series representative of Central England, originally compiled by Professor Gordon Manley, shows that since 1901 January has been colder than February 47 times, vice versa 45 times, while in two years the two months were equal.

Richard Inwards discovered that many countries on the continent had equivalent sayings, usually rather more colourful than the British one, indicating that the concept of warning each new generation not to be seduced by a few days of mild sunny weather in early February was general throughout Europe. From France we have:

> At the day of Candlemas
> Cold in air and snow on grass;
> If the sun then entice the bear from his den
> He turns round thrice and goes back in again.

And from Germany:

> On Candlemas Day the shepherd would rather have the wolf enter his fold than see the sun shine.

It is interesting that the Pilgrim Fathers took their weather lore with them across the Atlantic, for Inwards found this version in the United States:

> If the groundhog is sunning himself on 2nd February,
> He will return for four weeks to his winter quarters again.

BRAEMAR – BRITAIN'S COLD POLE?

As if to prove the point, the record for Britain's lowest ever authentically recorded temperature is shared by two dates, one before Candlemas Day and one after. A reading of −27.2 °C (−17 °F) was made on the morning 11 February 1895 and equalled early on 10 January 1982, and on both occasions the record was set at the Aberdeenshire village of Braemar, which can thus legitimately claim to be Britain's coldest spot.

Braemar (population 350) is located some 50 miles west-southwest of Aberdeen, nestling (as they say) in the valley of the River Dee, deep in the Grampian Mountains. Actually, to be strictly accurate, it sits astride Clunie Water, which is a tributary of the Dee, less than a mile south of the confluence of the two streams. The valley floor here is some 1100 feet above sea level, while Lochnagar (3768 feet) lies seven miles to the southeast, Ben Avon (3843 feet) seven miles to the north, and Ben Macdhui (4296 feet) eleven miles to the northwest.

After sunset, the ground radiates heat energy to outer space, but of course there is no warmth from the sun to replace it. That is, in the simplest possible terms, why it gets cold at night. The ground itself gets cold first, and the cold ground then cools the air in contact with it. A cloud cover impedes that loss of energy, while winds will mix the developing cold layer near the ground with warmer air above. If there is no cloud the warmth will continue to escape throughout the night, and if there is no wind the layer of cold air near the ground will become progressively colder and deeper. That is why clear, calm nights are usually coldest – in fact, they are known in the business as "radiation" nights. Cold air is denser than warm air, and on radiation nights it will tend to sink almost imperceptibly down the hillsides and into the valley bottoms, from where it will flow gently down the valley, eventually reaching broad lowlands where it will collect in a sort of "lake" of cold air. This process is known as "katabatic drainage", from the Greek *kata-*, meaning "down" or "away", and *bainein*, meaning "to go".

Braemar fits the bill perfectly. Practically surrounded by high mountains, an almost inexhaustible supply of cold air will find its way into the two valleys, and in the midst of a mid-winter cold snap the few hours of low, slanting sunshine during the day are quite insufficient to break up the pool of cold dense air. The conjunction of Glen Clunie and the Dee valley means that there is also a sort of aerial traffic jam as the two separate cold flows meet – the valley downwind of the village actually narrows – and the air in the Braemar bowl become virtually stagnant.

There is one further important additional point. The night-time drop in temperature is accentuated when there is a snow cover, particularly if the snow is deep and fresh and powdery. As the temperature at the earth's surface drops, some of the cold will be carried downwards into the ground by conduction. (If you stir the peas cooking on the stove with a metal fork, the fork handle will eventually become hot – the heat is being conducted from from one end of the fork to the other. A metal fork will get hotter quicker than a wooden or plastic fork, because metal is a good conductor of heat, but wood and plastic are not.) Bare soil is a fairly poor conductor of heat or cold, but fresh snow is 10 to 20 times worse, so there is a very rapid build-up of cold in the surface layer of the snow cover, which is immediately transmitted to the air in contact with the snow. Braemar has an average of 60 days snow cover per winter, and on the surrounding slopes that average is over 100.

A weather-recording station was first established in this Aberdeenshire village in 1855 under the auspices of Prince Albert, at about the time that Balmoral Castle (eight miles away down the A93) was built. In 40 of the last 100 years Braemar has recorded the lowest officially recognized temperature in the British Isles, and it also holds the record for 22 of the 90 individual dates in December, January and February. During the 1982 freeze-up the temperature fell to −22.6 °C (−8.7 °F) overnight on 6/7 January, −24.9 °C (−12.8 °F) on 7/8 January, −24.2 °C (−11.2 °F) on 8/9 January, −27.2 °C (−17.0 °F) on 9/10 January – equalling the all-time record, and −26.3 °C (−15.3 °F) on 10/11 January. During daylight hours on the 10th the temperature climbed no higher than −19.1 °C (−2.4 °F), the lowest afternoon maximum temperature ever recorded in Britain. During the winter of 1895, readings below −20 °C (−4 °F) were obtained early on 9 January and subsequently on the mornings of 8, 9, 10, 11, 13, 17, 18, 19, and 20 February. Both spells were characterized by deep powdery snow.

To be fair to this austerely beautiful part of Scotland, it is not always an Arctic outpost. The geographical features that contribute to the pooling of cold air on calm winter nights also encourage high temperatures on calm summer days. The temperature reaches 26 °C (79 °F) once a year on average, and has reached 30 °C (86 °F) on two occasions – 11 August 1975 and 8 July 1976.

MORE BAD NEWS FOR BRASS MONKEYS

We are very lucky to have such a long, virtually uninterrupted record from such an interesting location as Braemar. But there are only 500 or so official temperature-recording sites around the country, so it is highly improbable that Braemar was really the coldest spot in the country on either 11 February 1895 or 10 January 1982. After January 1982 it was possible to get some idea of areas that may have been colder, by assessing the degree to which local vegetation had been blasted by the severe cold. One area that appeared to be worse hit than the village itself was around Auchallater, some two miles to the south, where the valleys of Clunie Water and Callater Burn converge.

A similar effect is sometimes demonstrated in dramatic fashion by sharp late-spring frosts. A spectacular instance of this occurred on 17 May 1935 in the Home Counties. The temperature that morning fell to −8.3 °C (17 °F) in the notorious Hertfordshire frost hollow near Rickmansworth, but serious frost damage was confined to the lower parts of this particular valley. This state of affairs was mirrored in several neighbouring valleys in west Hertfordshire and south Buckinghamshire. One observer noted that the upper boundary of frost-blasted vegetation was so sharp, "it was as if the 400-foot contour had been etched onto the countryside by some sort of celestial cartographer".

Nowadays, satellite images can be analysed to reveal a great deal of temperature information at the time of each satellite pass. The temperature values are essentially averages for squares roughly half a mile across on the earth's surface, so they are not strictly comparable with readings taken from thermometers in Stevenson screens. What they do provide, though, is an incredibly detailed view of the geographical distribution of low temperatures, enabling us to identify all those places in Britain that could turn out to be colder than Braemar. But Braemar's record is in no immediate danger. It would need someone to establish official weather stations in all these cold areas, and then to record temperatures every day until one of those rare spells of intense cold turned up.

One such study was carried out by the Meteorological Office on satellite pictures taken during the early hours of both 10 and 14 January 1982. Many of the cold areas highlighted by this exercise were already well known, such as the broad middle section of Strathspey, roughly between Grantown-on-Spey and Aviemore. A few areas were quite a surprise, such as the northern fringes of Bristol and the western and northern outskirts of Wolverhampton. There was also a third group – and these areas were shown to be far more extensive than previously believed. For instance, the Shrewsbury district is well known for several cold pockets, but satellite analysis identified a huge cold tract covering much of northern Shropshire together with adjacent parts of southwest Staffordshire and east Montgomeryshire. The authors of the study considered that several areas in the Welsh Marches and parts of mid-Wales probably match the Dee and Spey valleys for extremes of cold, with all these areas at risk from very rare readings between −27 and −30 °C (−17 and −22 °F).

Apart from a few hardy eccentrics who are excited by extremes of all kinds, most people are allergic to this degree of cold. There are alternatives. There are various parts of the country that are disinclined to drop below −10 °C (14 °F) even under the most favourable conditions. These include hill tops, exposed sections of coastline, small offshore islands, and city centres. There is always a down side, of course: hills, coasts and islands are often windy and therefore a large wind-chill effect sometimes has to be taken into account, although the temperature according to the thermometer may not be particularly low; while city centres are noisy and polluted (and usually expensive to live in).

The Chiltern Hills provide some excellent examples of warm hill tops as well as some infamous frost hollows. During that severe spell in January 1982, a minimum of −19.7 °C (−3.5 °F) was recorded at an amateur station in Great Gaddesden at the bottom of the Gade valley just north of Hemel Hempstead, and −21.0 °C (−5.8 °F) at Wallingford on the Thames. By contrast, Whipsnade, some 770 feet above sea level, fell no lower than −8.9 °C (16.0 °F). There were even larger differences in the northwest Midlands where Newport in Shropshire (211 feet above sea level) recorded England's lowest ever temperature of −26.1 °C (−15.0 °F), while at Keele University, some 18 miles to the north and at an altitude of 590 feet, the minimum was −9.0 °C (15.8 °F). Hampstead, benefiting both from its hill-top site and its urban surroundings, logged a lowest temperature of only −6.5 °C (20.3 °F).

As for coasts and islands, Falmouth's minimum for the entire winter of 1981–82 was −1.5 °C (29.3 °F), St Mary's in the Isles of Scilly recorded nothing below 0.8 °C (33.4 °F) during December 1981, while on the Cornish mainland Delabole near Tintagel dropped to −8.7 °C (16.3 °F); no figures are available for St Mary's for January 1982. During a hundred years of records, St Mary's all-time low was −5.0 °C (23.0 °F) – exactly the same as Orlando in Florida and Palma in Majorca.

OUR COLDEST FEBRUARYS

"As the days grow longer, the cold grows stronger." So runs the old saying, and February regularly provides evidence to support the centuries-old weather lore. Long-term temperature records show that, on average, January is the coldest month of the year over the greater part of the British Isles, although February is fractionally colder in some western and southwestern districts where maritime influences are dominant – for instance, in Cornwall, Devon, the Channel Islands, Dyfed, Anglesey, the Isle of Man, and some of the Scottish islands. The North Atlantic is at its coldest in late February and early March – the relatively large thermal capacity of water means that an extensive area of sea warms up and cools down very slowly, and this is reflected in the eight- to ten-week time lag between the solstices and the warmest and coldest periods.

Professor Manley's Central England Temperatures series, culled from a variety of sources, and as far as possible standardized to ensure homogeneity, extends back to 1659. Over the whole series, February is 0.6 °C warmer than January and 0.3 °C colder than December, but this hides the fact that January has been coldest 44 per cent of the time, February 26 per cent, December 22 per cent, March 5 per cent, and November 3 per cent. Looking at decades, February was the coldest month of the year during the 1850s, the 1920s, the 1950s, and the 1980s. The 1860s was an extraordinary decade, for both December and January were colder than February, while March had almost exactly the same mean temperature, and November was only 0.6 °C warmer.

However, during the last hundred years, intensely cold Februarys have outnumbered similarly cold Januarys five to three. February 1947 averaged −1.9 °C, February 1895 −1.8 °C, February 1986 −1.0 °C, February 1963 −0.7 °C, and February 1956 0.2 °C. The only Januarys were 1963 with −2.1 °C, 1940 with −1.4 °C, and 1979 with −0.4 °C. We have had no December to match these freezing months since 1890.

One of the reasons for this surfeit of icy Februarys appears to be that "blocked" weather patterns, favouring cold northerly and easterly winds, are more frequent and longer-lasting in February, and this is partly a result of the decreasing vigour of Atlantic weather systems during the late-winter period. Very cold Januarys usually have one or two westerly intrusions, although January 1963 was a notable exception to this.

February 1986 was arguably the best example of a completely "blocked" month. Winds from an easterly quarter blew across England on 23 days during the month, the remaining 5 days having very light breezes of indeterminate direction. The temperature probably remained below freezing throughout the entire month above an altitude of about 1800 feet in the Pennines and Welsh hills, and the highest value of the month even at Hampstead (altitude 450 feet) in north London was a mere 2.9 °C. Snow covered the ground from the afternoon of the 5th onwards over a large part of eastern and southern England, while large areas in the west – sheltered from the prevailing easterly winds by mountains – had no measurable precipitation whatever. The sun shone for only 41 hours at Cupar (Fife) near the east coast of Scotland, but for 144 hours at RAF Valley (Anglesey) in the lee of Snowdonia.

It may come as a surprise to hardy northerners, but it is a fact that intense cold is a more frequent visitor to the towns and cities of the South and Midlands than it is to heavily populated areas further north. Southerners may or may not be soft, but it is true that Oxford and Cambridge endure longer sub-zero spells than, say, Manchester or Glasgow.

The record-breaking cold spell in January 1987 came to us courtesy of an easterly airstream originating over northern Russia, just this side of the Urals. The adjective "Siberian" seems apt to describe such weather but in truth air masses rarely if ever reach this country from the far side of the Ural mountains. A search through the records reveals that this 1987 easterly outbreak was probably the coldest this century. On 12 January, the temperature climbed no higher than −6.0 °C even in central London (St James's Park), −7.0 °C at Cheltenham, −7.7 °C at Herstmonceux observatory in East Sussex, −9.1 °C at Tonbridge in Kent and −9.2 °C at Warlingham on the southern outskirts of London. Further north, a comparatively temperate Manchester and Glasgow both managed −5.0 °C, while Edinburgh and Aberdeen reached −2.5 °C.

London's longest unbroken stretch below freezing point occurred during the infamous February of 1947, when the Royal Observatory at Greenwich remained sub-zero for 14 days. Neither Manchester nor Glasgow nor Aberdeen nor Belfast can remotely match such a long freeze-up.

THE NORTH SEA FLOODS OF 1953

Someone once said, "There is no such thing as a natural disaster, there are only natural events; the social and commercial activities of mankind upon which the natural event may impinge are responsible for any disasters." The logic may be unimpeachable, but it is of scant comfort to anyone whose life is directly affected by one of these catastrophic events.

Under the cloak of darkness, one of the 20th century's worst natural disasters in Europe stole in with frighteningly little warning on the night of 31 January – 1 February 1953 when a violent northerly gale brought a massive tidal surge down the length of the North Sea. On this occasion almost 100,000 lives were directly affected, and around 2000 lives were lost – the majority of them in the Netherlands. The figures for the United Kingdom were 307 drowned and 32,000 temporarily homeless. In eastern England roughly 300 square miles of land was flooded, while in the Netherlands according to some accounts an area of about 2500 square miles was under water – over one-sixth of the nation's territory. Livestock losses were put at almost 300,000. Total insurance losses for both sides of the North Sea were put at approximately £1000 million at the time, which translates to about £15,000 million at 1994 prices. The Dutch had to race against time to seal the hundred-odd breaches in the dikes before the following winter; the task was completed in November.

The disaster was caused by the coincidence of a variety of factors, mainly meteorological and oceanographic but also geomorphological and geographical. The storm surge in the North Sea was induced by a vigorous Atlantic depression tracking eastwards just to the north of Scotland, and then southeastwards into the North Sea. As soon as the wind veers into the northwest across the gap between northern Scotland and southern Norway, water is forced from the Atlantic Ocean into the North Sea. North Sea surges are particular vicious creatures because the sea is a comparatively shallow one, and it becomes progressively shallower and narrower towards the south, especially from Norfolk onwards, so the effect of the additional water is magnified dramatically in its southern portion. This particular surge was estimated to have raised the level of the southern half of the North Sea by an average of seven feet.

On this occasion, the wind continued to veer as the depression centre tracked towards Denmark, and the pile-up of water continued for almost 48 hours. The "low" was at its most intense over the central North Sea at about midday on 31 January, and it appeared to stall there for some hours before resuming its eastward progress, now gradually losing intensity. In its wake, pressure rose dramatically as a large high-pressure system developed to the west of Scotland. The temporary stalling of the depression while pressure was recovering over Scotland led to a temporary tightening of the pressure gradient to unprecedented levels over Orkney, Shetland, eastern Scotland, and the western North Sea, and a peak gust of 125 mph was recorded at Costa Hill in Orkney. The long fetch of storm-force northerlies along the length of the North Sea resulted in a build-up of a huge swell, which in turn produced massive breaking waves along exposed sections of coastline. A final meteorological factor was the low pressure itself

– a drop in pressure of 35 millibars (one inch of mercury) over the surface of the sea causes that surface to rise by about a foot.

The main oceanographic factors, apart from the development of the wind-generated swell, were the timing and the height of the astronomically induced tides. Ordinary tides depend upon the gravitational pull of the sun and moon on the oceans of the earth. Around British coasts there are two tides per day, and for two phases during each month the sun and moon are pulling sufficiently together to create "spring" tides, which are much higher than the intervening "neap" tides. The height of spring tides varies according to the precise alignment of the earth, moon and sun, and this one was several feet lower than the maximum possible. The surge on the night of 31 January – 1 February 1953 coincided almost exactly with high water along several sections of the English coastline, although in the Thames estuary it came two to three hours before (high tides in the Thames estuary are delayed in comparison with the rest of the southern North Sea as a result of the complex tidal geometry of the area). This difference diminished the surge heading westwards up the Thames, and saved central London from serious flooding.

The one important geomorphological factor concerns the very gradual sinking of the southern North Sea, which has amounted to some 15 feet since Roman times and continues to the present time. This means that the older sea defences very slowly become less effective, and overtopping during storm surges is likely to occur more frequently.

In the wake of this catastrophic storm surge, Winston Churchill, then Prime Minister, insisted that the sea would never again be allowed to launch such an effective invasion, and initiated a rolling programme of improvements to sea defences. This effectively set government policy for forty years, culminating in the successful construction of the Thames Barrier, completed in 1982. These improvements saved eastern England from a potentially worse disaster in January 1978, when a similar northerly storm coincided with a spring tide. But the sea has made occasional incursions to other parts of the country: serious flooding hit Lancashire's Fylde peninsula in November 1977, while the small coastal town of Towyn in northeast Wales was completely inundated after the breaching of a sea wall in February 1990.

It was not until 1993, when agriculture minister Gillian Shephard questioned the cost of defending agricultural land against the sea in an era of over-production, that a broad change in government policy was signalled.

"FEBRUARY, FILL DYKE" FAILURES

Over 80 per cent of us live in towns and cities, and the idea of imploring the rain gods to deposit yet more water upon us after the long, dark, damp days of a British winter may seem particularly perverse. But for a farmer, especially in eastern England, the exhortation is simply common sense, as explained earlier in this chapter. However, the prayer for regular rainfall inherent in the "February, fill dyke" weather lore often remains unanswered, for February is one of the drier months of the year, particularly in those eastern counties.

As a very rough rule of thumb most of lowland Britain averages 2 inches of rain per month throughout the year. Normal rainfall for February ranges from just 1.2 inches in a few spots in Essex and Cambridgeshire to 5.2 inches at Fort William and to 16 inches at The Stye (the wettest place in the UK for which rainfall records exist), high above Borrowdale in the Lake District.

Returning to eastern England, rainfall records for the Spalding district of Lincolnshire, where the average February rainfall is 1.5 inches, extend back over two and a half centuries. Since 1726 there have been 92 Februarys with under one inch of rain, and 25 with half an inch or less. Put another way, about one-third of all Februarys have less than an inch of rain, while almost one in ten have less than half an inch. These figures are crucial because in an average February evaporation accounts for a loss of moisture from the ground equivalent to half an inch of rainfall, and in a bright breezy month it may be as much as an inch. If evaporation exceeds rainfall, the land will begin to dry out, and that is almost the last thing a Lincolnshire farmer needs before those droughty March winds begin to blow.

Arguably the driest month since rainfall recording began was the February of 1891. Averaged over the whole of England and Wales, that month's rainfall totalled just 0.14 inches, barely one-twentieth of the long-term mean, and large tracts of the country remained quite rainless throughout. Ranking second in the list of dry months is June 1925 (0.17 inches), followed by January 1766 (0.18 inches).

The winter of 1890–91 must have been one of the strangest ever known. Severe wintry weather set in unusually early – in the last week of November – and lasted until mid-January, leaving in its wake the coldest and dullest December on record. February was, as we have seen, unprecedentedly dry and also very foggy, although several days of sunshine in the last week saw temperatures climbing to 19.4 °C (67 °F) at Llandudno on the and Cambridge on the 27th, a record for February that has stood ever since, but this enchanting early taste of spring gave way to a ferocious snowstorm in early March.

By delving into the meteorological journals of the time, one can get a flavour of this quite extraordinary February. One rainfall observer, a Mr C.A. Case, tells us that, near Maidstone, "A field with long grass took fire from a man's pipe-light, the whole of the grass was burnt off, and some of the quick-set hedge burnt and destroyed; accounted for by the dry month and the frost having previously killed the grass." Mr W. Lucas, the observer at Hitchin, Hertfordshire, wrote, "On 1 February enough rain fell to make 0.01 inch and so spoil my carte blanche. This month's maximum temperature is unprecedented, 62 °F. Sulphur butterflies have swarmed during the last fortnight and on the 16th a queen wasp was killed. Altogether, everything is out of gear . . ."

February 1891's weather was entirely dominated by high-pressure systems, which covered the British Isles on practically every day during the month. But small amounts of rain did fall around the periphery of these anticyclones from time to time, and the western and northern parts of Scotland therefore had rather more rain than the rest of the country.

Sometimes the normal distribution of rainfall is completely reversed, with comparatively large amounts in southern and eastern England, and very little in the western Highlands. Such a month is the result of long periods of easterly winds, such as the Februarys of 1947, 1963, and 1986. Each of these months was bitterly cold, and snow covered the ground practically throughout in some parts of England. Fort William recorded 0.08 inches in February 1947, no measurable rain during February 1963, and 0.10 inches in February 1986. The lack of rainfall was exacerbated by the desiccating east wind, and fires broke out on the heather-covered hillsides in parts of Inverness-shire and Wester Ross in each of these months.

Alongside the old adages about "February, fill dyke", there are several others that illustrate the fear of drought at this time of the year:

> All the months in all the year,
> Curse a fair Februeer.

And, from Cornwall, short and to the point:

> A February Spring
> Isn't worth a pin.

5. March

"March, black ram,
Comes in like a lion and goes out like a lamb."

ANON

Most lions that I've seen seem to spend most of their day dozing peacefully. But I guess they do have a considerable capacity for violence, and as such come to represent the violent winds that are alleged to be characteristic of early-March weather. The juxtaposition of lions and lambs to illustrate contrasting extremes certainly has a good pedigree, with a scattering of references in the Bible and indeed throughout English literature.

Thus we are meant to believe that a typical March begins with a spell of wild windswept days, and ends with pleasant spring sunshine. Common sense tells us that it doesn't really need a statistical analysis to prove that this is one of those sayings that have no predictive value; it probably provided a little colour to everyday exchanges in times past, and perhaps acted as one of those easy ice-breaking conversation-openers that we British seem to find necessary. For the record, a look at the 25 Marches between 1970 and 1994 revealed nine years with strong to gale-force winds affecting the greater part of the country on 1–2 March and just five years with high pressure in charge on 30–31 March. Only in 1982, 1989 and 1990 did March begin windy and finish sunny. Anyone who finds such statistics unconvincing may yet be swayed by the fact that collections of ancient weather lore include another old saying, reversing the one that introduces this chapter:

> March comes in like a lamb
> And goes out like a lion.

Although this second saying is less well known than the first, no one appears to know which, if either, was the original. Coincidentally, between 1970 and 1994 there were also just three Marches – 1976, 1980, and 1992 – that began calm and sunny, and ended with high winds. Perhaps the best way to look at this is to recognize that a typical March will contain contrasting spells of weather, both leonine and ovine, as befits a month that straddles the diffuse boundary between winter and spring.

Here we find another very important facet of March's weather. Broadly speaking, significant plant growth only takes place when the temperature is above 6 °C (43 °F), and that is very roughly the mean monthly temperature for March over large chunks of the British Isles. From place to place, therefore, mean March temperature will vary either side of that critical value, as it will from year to year in any particular location. This is why in an average March, spring will have sprung in the warmer parts of Britain by the month's end with the countryside clothed in green, while in colder areas winter appears to be hanging on remorselessly, the trees and hedgerows are bare, and daffodil buds remain resolutely closed. Similar contrasts will be apparent from one 31 March to the next in any one place.

The Central England Temperature series – comprising a mean of figures for lowland Lancashire and the middle Thames valley, and with records going back to 1659 – shows an average March temperature of 5.7 °C (42.3 °F) during the period 1961–90. The lowest average March temperature in the series was just 1.0 °C (33.8 °F) in 1674, and the highest 9.2 °C (48.6 °F) in 1957. One can imagine the startling contrast the English countryside must have displayed at the conclusion of those two months.

If, just for the sake of argument, we take that value of 6 °C as marking the threshold between a "winter" month and a "spring" month, the following interesting comparisons emerge. At the warm end of the country Plymouth (average March temperature 7.4 °C) has had 90 per cent of its Marches in the "spring" category and only 10 per cent in the "winter" category, while in the coldest corner at Aberdeen (March mean 4.5 °C) the figures are precisely reversed with 10 per cent of all Marches having been spring months while 90 per cent were winter months. The degree of overlap may be surprising, showing as it does that spring growth will be well advanced in late March in northeast Scotland in a handful of years, while in a few other years the normally mild West Country may reach the end of the month with little outward sign of spring.

Figures for other parts of the country are more evenly distributed. London scored 60 per cent warm and 40 per cent cold, Birmingham exactly 50–50, Belfast 40–60, Glasgow 37–63, and Newcastle 33–67.

All this just goes to show how sensible we are not to have an official first day of spring in Britain. Clearly, the start of the season varies hugely in time and in space. The land can spring into activity in January in Cornwall in a warm year, but not until late May in northeast Scotland in a cold one.

WHEN THE WIND BLOWS

"100 mph Hurricane Batters Britain" scream the headlines, not just in the tabloids either. But how was that wind-speed measured? Where was it measured? And does it really mean anything anyway?

Anemometers are the instruments that meteorologists use to measure both wind speed and wind direction. The word comes from *anemos*, the Greek word for wind. Rather more sophisticated than the simple anemometer is the anemograph, which automatically transfers the measurements onto a chart, thus producing a continuous record of the behaviour of the wind at a given place. In turn, these are being replaced for certain purposes by computerized systems that store the second-by-second variations in the wind onto disk, from which the forecaster or researcher can extract whichever details he needs.

There are two basic forms of instrument – the cup anemometer and the pressure-tube anemometer. The first works by means of the wind rotating an arrangement of hemispherical cups (usually three of them), which cause a verticle spindle to revolve, there being a near-constant ratio between the speed of the wind and the speed of rotation of the cups. The pressure-tube version relies on the change of air pressure in a pipe caused by the wind blowing across an opening at the end of the pipe.

The biggest problem in measuring wind is deciding where to site the anemometer. We, of course, have most of our houses and gardens and offices and factories and transport systems in the lowest 50 feet or so of the atmosphere, but this is precisely the area where winds are reduced and distorted by friction and turbulence. Friction reduces average wind speeds at head height by considerably more than 50 per cent when compared with winds in the free atmosphere several hundred feet above us. And you only need to walk the streets in any British town on a windy day to appreciate the degree of turbulence and funnelling caused by buildings.

The accepted standard for anemometer sites is thus a compromise – high enough to escape much of the distortion caused by buildings, trees, and so on (what meteorologists call "surface roughness"), but low enough to be relevant to the very shallow layer of the atmosphere in which we conduct our everyday lives. Thus most instruments are sited at an "effective height" of 33 feet above their surroundings. Even so, in a city environment the residual turbulence will still result in a gust ratio of more than two – that is, the strongest gusts are more than double the average wind speed.

One other important point has to be considered when

relating wind speed to gale damage. Essentially, wind speed is measured at a point but damage is caused by the force exerted by the wind over an area, so gale damage should in theory be roughly proportional to the square of the wind speed. In practice, the degree of damage will probably be exacerbated the longer the gale persists.

Reliable wind-speed records have now been kept for just over a century, but until the World War I there were comparatively few regularly working anemographs. By January 1914 there were about 20, mostly around the coastline, whose records were published by the Meteorological Office. In January 1940 there were still only 35 of them, and in 1960 just 68, but by 1990 the number had risen to 150. It is at least partly this proliferation of instruments in recent years – in particular the installation of heavy-duty ones on mountain tops – that has resulted in the apparent increase in the frequency of gust speeds of over 100 mph.

Up to a point, if you want to establish new wind-speed records, you can pick your place to erect your anemometer. The highest wind speeds are recorded on mountain summits and cliff tops. Thus, the United Kingdom's all-time gust-speed record of 173 mph was logged on the summit of Cairn Gorm (4084 feet above sea level) on 20 March 1986. But at the anemograph at Cairngorm Chairlift, some 500 feet below, the highest gust that day was 105 mph, while Aviemore in the valley below managed just 63 mph. Similarly the world record of 231 mph is claimed by the Mount Washington Observatory in New Hampshire, USA, on 12 April 1934. The summit of Mount Washington is 6288 feet above sea level.

A number of cliff-top sites in the UK have reported gusts in excess of 100 mph during recent years. Thus, after a particularly windswept day, news reports will be peppered with names such as Portland Bill, Brixham Coastguard, Gwennap Head, Butt of Lewis, and St Catherine's Point. Britain's low-level wind-speed record was established at Kinnaird Head lighthouse in Fraserburgh on the northeast coast of Scotland: on 13 February 1989 the wind there gusted to 142 mph.

Often as we see these locations quoted in the newspapers and on radio and television during windy weather, they have precious little relevance to you and me – there are of course very few trees to blow over or buildings to damage on the summit of Cairn Gorm. In a typical winter, very few gust speeds in excess of 80 mph are likely to occur in populated areas. This is why the violent winds recorded during the Great October Storm of 1987 in southeast England were so extraordinary. On that occasion, Shoreham in Sussex logged 113 mph, Langdon Bay near Dover 108 mph, Sheerness on the Isle of Sheppey 107 mph, and Ashford in Kent 106 mph. Two other stations in Sussex and one in Essex also exceeded 100 mph, while London Weather Centre in the heart of the capital recorded 94 mph. Weather forecasters use the 70 mph gust-speed mark as a very rough guide to indicate when winds are likely to uproot trees and cause localized structural damage.

MARCH HEATWAVES

Sometimes, the March winds stop blowing, the sun comes out, and the temperature soars. But March heatwaves are such rare creatures that memories of them should be cherished, for twenty or more years may pass from one to the next. Such a spell happened in mid-March 1990, with the temperature soaring into the low 20s C (low 70s F) over quite a large area extending from Surrey northwards to Yorkshire and westwards to Worcestershire. This was the first such occasion in March since 1968.

Such unseasonably high temperatures are sufficiently unusual to encourage us to investigate the mechanism responsible.

Broadly speaking, anticyclones – regions of high pressure on a weather map – bring us dry, settled weather, frequently with clear skies and sunshine. Certainly, a lengthy spell of settled, sunny weather requires a high-pressure system centred over or close to the United Kingdom. Under these conditions during the winter season (more specifically between early November and mid-February), with the sun above the horizon for only eight hours or so and relatively low in the sky even at midday, the incoming radiation from the sun exceeds the outgoing radiation for only a few hours, and therefore the temperature does not climb very far. Consequently, calm sunny winter days are also normally cold days.

Our warmest weather during the winter is "imported" – it is brought to us on southerly or southwesterly winds whose origin lies somewhere over North Africa or more usually over subtropical regions of the Atlantic Ocean. Over most of Britain, highest recorded temperatures in January are around 15 °C (59 °F), and occur under cloudy skies with quite strong winds, although a few favoured locations to the north of high ground enjoy rare days of 17–18 °C (63–65 °F) thanks to the "föhn" effect (see Chapter Three).

In summer, the incoming radiation during an anticyclonic spell exceeds the outgoing radiation for upwards of 12 hours daily, and calm sunny days become hot. But there are critical periods around the equinoxes when the radiation equation is much more evenly balanced, and it is at these times that other influences have a disproportionate effect.

An early or mid-March heatwave is most likely to develop when a large high-pressure system covers much of continental Europe, allowing southerly winds on its western flank to bring air of North African origin across Spain and France to eastern and central Britain. This combination of the anticyclonic and importation effects provides exactly the right recipe for abnormally high temperatures. But the "high" has both to have the right shape and to be in the right place to ensure exceptional warmth. Slight variations can shift the region of origin of the airflow to the Atlantic on one side, or to the Balkans on the other, both of which would result in maximum temperatures at least 5 °C lower. By late March, high temperatures may occur without the impetus of an airflow from a very warm source region.

March warm spells with afternoon temperatures into the 70s Fahrenheit (over 21 °C) have occurred in 10 years this century – in 1990, 1968, 1965, 1961, 1957, 1948, 1945, 1929, 1918 and 1907. The 1990 event seemed to be long overdue after the relatively frequent occurrences in the late 1950s and 1960s. During that long gap in the 1970s and 1980s many Marches seemed to be little more than an extension of winter, distinguishable from February only by the increasing length of daylight.

Undoubtedly, the most remarkable of these early heatwaves occurred in 1948 when, in the midst of a prolonged period with afternoon temperatures around 15 °C (59 °F) – warm enough for early March – the mercury soared into the 70s F on the 9th. Highest of all was 23.9 °C (75 °F) at Wealdstone in northwest London, but 23 °C (73 °F) was logged at a number of stations in London and the southeast and also in East Anglia, and 21 °C (70 °F) was touched as far north as Houghall near Durham and as far west as Totnes in Devon. This date, 9 March, remains the earliest date of the year on which 21 °C has been reached in Britain, and at a number of weather stations it is still the warmest March day on record. Logically one would assume that the highest temperatures in March would all occur towards the end of the month, making the 1948 event all the more noteworthy. Later in March 1948 exceptional temperatures occurred in Scotland, including 21.7 °C (71 °F) at Strathy, which is located near the northernmost extremity of the British mainland, just west of Thurso.

There were also two spells of outstandingly warm weather in March 1929. The widely quoted 25.0 °C (77 °F) at Wakefield on the 29th of that month is probably incorrect, being 5 °F higher than adjacent station reports (all temperatures were recorded in Fahrenheit until the 1960s). The next day 23.3 °C (74 °F) was authentically recorded at Killerton Park near Exeter, at Newport on the Isle of Wight, and at Kilkenny in Ireland. Earlier that month Keswick measured 22.2 °C (72 °F) on the 9th.

Rather more recently, mid-month heatwaves brought

23.3 °C (74 °F) to Haydon Bridge in Northumberland in 1957 and 22.8 °C (73 °F) to Goldington near Bedford in 1961. Wakefield tried again on 29 March 1965 with 25.0 °C (77 °F), and this value was equalled, also on 29 March, in 1968 at Santon Downham near Thetford, and at Cromer. These last figures represent the all-time record for March in the British Isles.

Just in case you wondered, March heat tells us nothing about the forthcoming summer. In 1990, a sequence of hot spells culminated in the heatwave of the century in early August, but in 1965 there were only three notably warm spells during the entire year – one in March, one in May, and one in October!

THE BLIZZARD OF '91

A slightly earlier heatwave saw temperatures soaring to 19.4 °C (67 °F) at the end of February 1891. Within two weeks, southwest England was manfully digging itself out of the Great West Country Blizzard of March 1891. This snowstorm ranks alongside those of January 1881, December 1927 and February 1978 as one of the worst since systematic weather recording began. Indeed, for over half a century, the Blizzard of '91 was a colloquial marker by which all subsequent snowfalls in the South-west were measured.

Some 220 people lost their lives – many of them at sea, for 65 ships were lost in the English Channel alone. In addition, over 6000 sheep perished, countless trees were uprooted, and 14 trains were stranded in Devon alone, most of them stuck in enormous snowdrifts. The Great Western Railway's "Zulu" express left Paddington at three o'clock on the afternoon of Tuesday 10 March, finally limping into Plymouth station at 8.30 in the evening on Friday the 13th. The accompanying northeasterly gale was so violent that on one farm sheep were simply blown over a cliff. These violent winds swept exposed fields practically clear of snow, while sunken lanes were buried to a depth of 20 feet or more. On the fringes of Dartmoor, the gorge-like valley of Tavey Cleave, some 300 feet deep, was reported to have been completely filled with snow. The main part of the blizzard lasted for 36 hours, from lunchtime on the 9th until the night of the 10th/11th, but snow and strong winds returned during the 12th and 13th.

Trains had to be abandoned in other parts of the country as well, and although the West Country was worst affected, southern England, the Midlands, and south Wales also suffered badly. Drifts were so huge around some houses in the London area that, according to a contemporary newspaper report, children amused themselves by jumping out of upstairs windows onto the snow.

This was probably the first time that the word "blizzard" had been used in this country. Curiously, the newspapers during the preceding few weeks had carried comprehensive reports of a sequence of severe blizzards in different parts of the USA, and it is quite conceivable that the word may not have crossed into popular parlance on this side of the Atlantic but for the coincidence of events – not until some later time, at any rate. The distinguishing features of a blizzard are the strength of the wind and the extreme cold, which allows dry powdery snow to be blown into drifts. In other words, the air is laden with snow, some of which at least has been lifted from the ground. The Americans have defined very specific criteria before a snowstorm can be upgraded to the blizzard category, namely a temperature of −7 °C or lower (below 20 °F), a mean wind speed of 30 mph, and horizontal visibility reduced below 500 yards. A severe blizzard requires the temperature to be below −12 °C (10 °F), a sustained wind of 40 mph, and zero visibility.

The meteorological circumstances responsible for the March 1891 blizzard were rather unusual. The warm sunny weather of February gave way to changeable westerly winds during the first week of March, then on the 7th a vigorous cold front swept southwards across the British Isles, introducing a cold northerly airflow. Some northerly outbreaks are very powerful, especially in early spring, and the Arctic air often penetrates into Spain and the western Mediterranean, and occasionally into North Africa. This one was much less vigorous and it ran out of steam just to the south of England. This left a well-defined boundary – a slow-moving cold front – between the cold northerlies and relatively warm and moist southwesterlies which blew across Spain, Biscay and France. Small ripples along the length of this cold front developed into active secondary depressions that tracked north-eastwards along the English Channel, keeping the whole of England in the cold northerly airflow. Precipitation was heaviest and the wind strongest near to the track of these depressions, which is why southwest England really caught it in the neck.

It has proved very difficult to reconstruct the distribution of snow depth at the end of this blizzard, because of the violence of the wind and the consequent severity of the drifting. The words of Mr M.A. Glanville, the rainfall observer at Ivybridge in south Devon, illustrate the frustration of the assiduous weather chronicler who had to leave his rainfall column blank for these days:

> My calculations for this month are much thrown out by the great storm of the 9th and 10th inst. The rain-gauge stood in a very few inches of snow; north-east of it, about thirty yards off, was a drift twenty feet deep; west of it, the snow increased in depth up to and above the entrance gate, about eight or nine feet.

As far as can reasonably be estimated, six inches of snow covered the whole of England and Wales south of a line from Caernarvonshire to Essex, and it was over a foot deep over the hills of south Wales, over the Cotswolds and the Downs of Berkshire and Hampshire, and over much of Cornwall and Devon. Except in parts of Cornwall and south Wales, therefore, less snow fell than in the January 1881 snowstorm. But in the West Country the gale was more violent and therefore more destructive, and the drifting more severe, and in the public consciousness in Cornwall and Devon the March 1891 blizzard was clearly the worse of the two.

THE WEATHER OF MARCH 1947 – EVERYTHING BUT THE KITCHEN SINK

Blizzards were also a feature of the winter of 1946–47. In fact, most weather historians regard this as the snowiest winter of the last 150 years. Mr Attlee and his colleagues must have thought that the weather had it in for his Labour administration. His nationalization programme was well under way, with the railways and the coal industry coming into public ownership, but there was considerable industrial unrest, which accentuated the problems caused by the severe shortages that followed the end of the Second World War. Fuel was one of the things that was in short supply even before the arctic weather began, but when the ice and snow arrived it became impossible to move sufficient fuel from the coalfields to the power stations, and most parts of the country endured daily power cuts during the periods of peak demand. Industrial activity plummeted, the level of unemployment soared, and the image of Buckingham Palace and the Houses of Parliament working by candlelight, less than two years after winning a war, was a telling one.

It was extraordinary that the really severe weather during this winter did not set in until the fourth week of January. And it was extraordinary that it continued through the first half of March, culminating in a series of highly disruptive snowstorms. During that first fortnight of March there was also a serious ice-storm in southern England, together with record-breaking frosts, occasional dense freezing fogs, and a couple of fierce easterly gales. During the second half of the month a rapid thaw and frequent downpours conspired to cause the worst river floods ever, and a violent westerly gale battered southern Britain on the 16th. Hail and thunder were also noted, and it seemed that the only sort of weather that failed to appear was one of those rare March heatwaves – the highest temperature anywhere in Britain in March 1947 was just 15.6 °C (60 °F).

The effect of this wicked winter on agriculture was even

worse than the effect on industry, and much longer lasting too. Cereal production in 1947 was 10 per cent down on the previous year, and potato production dropped by 15 per cent. Wheat yields averaged 0.95 tons per acre in 1945 and 1946, dropping to 0.78 tons per acre in 1947, but had recovered to 1.12 tons per acre by 1949. Almost a quarter of Britain's sheep were lost, and the 1946 total was not exceeded until 1952.

As the month opened, much of the country had been continuously snowbound for over five weeks, and the snow was deepest over eastern counties from Norfolk up to Aberdeenshire, especially over the hills. At Lincoln level snow was 10 inches deep, at Huddersfield 17 inches, and at Ushaw, near Durham, 28 inches.

Put simply, March 1947's weather was caused by a sequence of seven large Atlantic depressions that tracked northeastwards in the vicinity of the British Isles. The detail of the weather depended on the track of each low, and each of the first six travelled on a course slightly further north than that of its predecessor. Consequently, the snowstorms on the cold northern side of the depression tracks swept areas progressively further north, while thaws on the warm southern flank affected a larger area each time.

Depression number one passed up the English Channel between the 4th and 6th, but in the lull before the storm the country was deep-frozen, with the month's lowest temperature of −21.1 °C (−6 °F) being notched up at Braemar, Peebles, and Houghall (near Durham). At the time this was a new all-comers record for March, although even colder March nights subsequently occurred in 1958 and 1965. The snowstorm raged for 48 hours over practically the whole of England and Wales, completely disrupting rail and road traffic, and even London was totally paralysed. South of the Thames the snow turned to freezing rain, resulting in a bad ice-storm across parts of Kent, Surrey, Sussex and Hampshire. Further north, level snow was 16 inches deep at Edgbaston (Birmingham), 3 feet deep at Lake Vyrnwy in mid-Wales and 5 feet deep at Clawdd-newydd, near Ruthin in north Wales. There were drifts 9 feet high in the Chiltern Hills and 16 feet high in the Brynmawr–Tredegar–Ebbw Vale district in northwest Gwent. In the wake of this storm the temperature plunged again to −21.1 °C (−6 °F) at Braemar and to −18.3 °C (−1 °F) as far south as Droitwich, Worcestershire.

Depression number two raced across southern England on the 11th, and number three tracked across central Ireland, Wales and the Midlands during the 13th and 14th. This time, northern England and Scotland endured blizzard conditions, and average snow depth reached 34 inches at Ushaw and 18 inches at Harrogate, with drifts up to 25 feet high north of the border. Meanwhile temperatures in southern counties jumped dramatically, reaching 10 °C (50 °F) for the first time in eight weeks, initiating a rapid thaw and widespread floods.

The fourth depression hurtled northeastwards across Wales and northern England on the 16th, and, as if to rub salt into the wound, a violent gale battered the southern half of the country that evening. At Cardington, near Bedford, the wind averaged 63 mph and gusted to 93 mph at the height of the gale, while Mildenhall in Suffolk reported a gust of 98 mph. At the same time milder air was finally introduced to most of Scotland, albeit briefly. Numbers five, six and seven brought mostly rain, but this was now a serious problem in its own right, and the repeated steady downpours during the second half of the month exacerbated the widespread flooding caused by the thaw. Indeed March 1947 was easily the wettest March in 250 years of records over England and Wales. Large parts of the country had over three times their normal rainfall, and Torquay had 371 per cent of the local average. Meanwhile sunshine was very deficient, both Wakefield and Bolton totalling just 46 hours – barely half the long-term average.

THE SPRING EQUINOX

The equinoxes occur on or about 21 March and 22 September. The spring (or vernal) equinox in the northern hemisphere varies between 20 and 22 March, according to our position in the leap-year cycle; similarly the autumn equinox occurs on 21 or 22 or 23 September.

The word derives from the Latin *aequi* meaning "equal" and *nox* meaning "night", for these are the two dates during the year when – in theory – day and night are of equal length the world over, with the sun everywhere above the horizon for 12 hours. In practice, this is not quite true. Local geography does, of course, cause irregularities in the horizon in any particular place, and the fact that the earth is not precisely spherical also has a tiny effect. But the main influence on the theoretical daylength is the fact that the sun's rays are refracted (that is, slightly bent) as they pass through the earth's atmosphere. In the latitude of the British Isles this adds something between six and seven minutes' daylight. In other words, we see the sun rising just over three minutes before it is actually there, geometrically speaking, and we see it setting just over three minutes after it has actually gone. This leads to a curious paradox at the poles. At the North Pole, following the six-month-long winter night, the sun should in theory travel around the whole horizon on the date of the equinox, thereafter climbing a little higher each day at the beginning of the six-month-long summer day. But because of the refraction effect the sun will be just above the horizon throughout both equinoxes at both poles.

Some people still like to call the spring equinox the beginning of spring, and many are surprised to learn that there is no such thing as an official first day of spring in Britain, and there never has been. The idea of four equal seasons separated by the equinoxes and the solstices is a purely astronomical one and has little meteorological or botanical relevance. The climate historian E.L. Hawke believed that this astronomical division dates back to an ancient Greek (or more properly Rhodean) astronomer called Geminus, who set forth his ideas around 70 BC. The scheme quickly fell into obscurity, but resurfaced again in the USA early in the 19th century, and to this day Americans celebrate 21 March as the first day of spring with near-reverence. British Victorian almanacs copied the idea, and some diaries maintain the notion. A rather more logical arrangement would be to adopt the equinoxes and solstices as the mid-points of the seasons – with the longest day at the mid-point of summer, for instance – but nobody has made any serious attempt to do this.

Here in Britain, the keepers of official weather records regard spring as comprising the months of March, April and May, so for them 1 March is the first day of the season. This is the nearest we come to anything "official", but in truth it is merely a matter of convenience – it simplifies book-keeping if each season consists of three calendar months.

In the distant past, country folk frequently counted February, March and April as the three spring months, since February was the time when (to coin a phrase) the first green shoots usually appeared. Another ancient definition owes much to the close association of church and country life over hundreds of years, and our forebears often associated weather events with saints' days. According to Hawke, the scheme was described as follows:

> Dat Clemens hiemem; dat Petrus ver Cathedratus; Æstuat Urbanus; Autumnat Bartholomaeus.

Thus winter was deemed to begin on St Clement's Day (23 November), spring on St Peter's Day (22 February), summer on St Urban's Day (25 May) and autumn on the feast of St Bartholomew (24 August). It may be significant that the four Quarter Days – the dates at which rents, tithes, and some other taxes were due – occur almost exactly a month after each of these days.

Another late-March event that marks the advancing season is the great British celebration of remembering to the put the clocks forward at the beginning of British Summer Time. Daylight saving has been with us in one form or another for over three-quarters of a century. First introduced in wartime Britain in 1916 as an energy-saving measure, it is difficult now to appreciate the vehemence with which it was criticized. Some of the criticism was well argued but much of it was jingoistic, even in learned circles.

One scientific journal called it variously "sham time" and "Prussian time", while Dr Hugh Robert Mill, a leading meteorologist, offered some heavy-handed sarcasm: "No doubt Parliament in compelling us to keep the time of the enemy meridian does not really expect us to believe that German time is our time. . .this involves the same abominable principle that would promote temperance and attendance at public worship by providing that on Sunday St Paul's shall be called the Red Lion, and on weekdays the Red Lion shall be called St Paul's." But another correspondent put the complaints into some sort of context by reminding readers that "the fact that noon is now that of Berlin need not cause the patriot to shudder at the change because in Germany and its subject lands the noon of Petrograd is used to set the summer clock."

Between 1941 and 1945 inclusive, and also in 1947, double summer time was in force, with GMT+1 in winter and GMT+2 in summer, while the first experiment with British Standard Time (GMT+1 throughout the year) lasted from 1968 to 1971.

6. April

"March winds and April showers
Bring forth May flowers."

ANON

April may be famous for its showers, but it is rarely a very wet month. Quite the contrary; April is one of the drier months of the year in all parts of the British Isles, and in some parts it is the driest of all. The record books contain several examples of Aprils which have been completely rainless over quite large tracts of the country. The most recent example was in 1984, when no measurable rain fell in the neighbourhood of Oxford, Portsmouth, Bournemouth and Swindon. Other remarkably dry Aprils in England and Wales include 1957, when Hereford, Malvern, Monmouth and Bridgwater were rainless; 1938, when a large area around Bournemouth again escaped rain; and 1912, when large parts of the London area had none. In April 1974, parts of Wester Ross in the far northwest of Scotland were completely dry.

So why should April have this reputation of being showery? It is certainly not that showers are more frequent in April than at other times of the year – far from it – but over inland parts of England this is the season when showery activity increases rapidly. Unlike sunny mornings in February and early March, which usually herald fine days, many sunny mornings in April and successive months flatter to deceive. That familiar build-up of cumulus clouds during the morning, blotting the sun out for short periods, is so frequently followed by sudden sharp showers from late morning or early afternoon onwards.

Let us first define what weather experts mean by a shower, for it does not always coincide with the use of the word in popular parlance. The *Observer's Handbook* states:

> Showers always fall from convection clouds (that is, cumulus or cumulonimbus). The amount of cloud usually varies greatly during the course of an hour or so, and often within much briefer periods. Shower clouds may usually be seen either building up before the shower begins to fall, or approaching before the precipitation arrives. When a well-developed shower cloud is over a location it may for a time cover the whole of the sky, but after the shower there is usually a partial, and sometimes a complete, clearance of the sky; the cloud may be seen to decay or to move away after the precipitation has ceased. Hail is always a shower type of precipitation.

Cumulus clouds, starting off like small scraps of dirty cotton wool, developing into larger "cauliflower" clouds, will under suitable conditions eventually build into towering cumulonimbus. They are associated with convective currents, sometimes called thermals, which are columns of warm rising air that can extend from the earth's surface to anything up to eight miles above it. The more vigorous the convection, the more intense the showery activity becomes, with hail and thunder increasingly likely.

These convective currents develop when warm surface air is overlain by colder air aloft. A bubble or pocket of warm air rising from ground level will become more buoyant as it enters the colder layers, and may continue rising until it reaches the base of the stratosphere, where the temperature structure changes. Such a contrast between surface and upper air frequently happens when a cool northwesterly or northerly airflow originating in the Greenland region or over the Arctic Ocean covers Britain, and strong morning sunshine brings a rapid temperature rise in the lowest few feet of the atmosphere. This is a common weather pattern during April.

The most violent convective activity seen in this country occurs during the summer months. This can take the form of severe thunderstorms at the end of a hot spell when cool and moist Atlantic air overlays hot continental air. Or it can occur during the summer equivalent of the Arctic-air outbreak, with unusually cold air of polar origin in the upper atmosphere and powerful early sunshine warming up the surface layers. Such a combination can result in those dramatic storms when a hail-covered summer countryside takes on a Christmas-card appearance.

Many weather forecasters have a peculiar blind spot about using the word "showers" in their forecasts. In general conversation, showers are normally light sprinkles of rain (whether or not from convection clouds), but your average forecaster sticks religiously to his or her official definitions. This is fine for communicating with other meteorologists, but it leads to a regrettable lack of communication with all those thousands of ordinary people reading or listening to the forecast. It is amusing to see the perplexity on the face of a typical met-man, who, pleased with himself for having correctly predicted heavy showers, is laughed at by his non-expert companion: "That was no shower, that was a ruddy cloudburst!"

WINTER ARRIVES . . . IN APRIL

April showers or no, it is natural to think of spring as the season when the weather gets progressively better, with temperatures rising a few degrees week by week. In practice, few if any British springs conform to this idealized concept; typically, warmer weather arrives in fits and starts, and setbacks can be quite dramatic.

Sudden sharp frosts in April and May have been the bane of fruit and vegetable growers since time immemorial, and hill farmers similarly dread heavy April snows, which can exact a heavy toll of the season's new-born lambs. For the rest of us such uncomfortable reminders of winter are usually no more than an irritation, upsetting our work in the garden, or making the first cricket matches of the season an ordeal rather than a pleasure.

A dramatic sunbathing-to-snowshoes flip in the weather in March or April provides good copy for the newspapers, and headlines frequently refer to the climate going crazy. But there is nothing abnormal or sinister about such wide meteorological fluctuations at this time of the year. Put simplistically, it is just Mother Nature's way of balancing the books. Spring is the season when the prevailing southwesterly winds in temperate latitudes are at their weakest. The jet stream in the upper atmosphere meanders first polewards then equatorwards, and as a consequence winds near ground level are often northerly and southerly in adjacent sectors of the hemisphere. It only needs a slight shift of the jet stream for the low-level wind in a particular area to switch from southerly to northerly or vice versa, and air of tropical origin is suddenly replaced by air of Arctic origin, with all that that implies in terms of weather.

One of the most breathtaking of such reversals occurred in 1968. On 29 March that year, the all-time record March temperature was equalled with 25.0 °C (77 °F) at Cromer and at Santon Downham, near Thetford, but early on the morning of 2 April a reading of −10.6 °C (13 °F) was logged at Achnagoichan in Strathspey. By midday central and eastern England had a covering of snow with temperatures close to freezing point and a penetrating north wind blowing, and a few days later an early-morning minimum of −9.4 °C (15 °F) was recorded at Caldecott, on the Leicestershire–Northamptonshire border, near Corby.

Similar, though rather less dramatic, sequences of events happened in late March and early April in both 1989 and 1990. March 1989 ended with a week of bright warm weather with

20.7 °C (69.3 °F) at Elmstone, near Canterbury, on the 28th and 19.9 °C (67.8 °F) at the London Weather Centre on the 31st. It was warm and sunny again on 1 April, but subsequently cold east winds set in, bringing heavy snow to London and the Home Counties on the morning of 5 April – at midday level snow lay 6 inches deep at Tadworth, high in the North Downs on the southern fringe of Greater London. Further north, there was a 16-inch-deep cover at Holme Moss (West Yorkshire), located high above Holmfirth in *Last of the Summer Wine* country. I suspect filming that day was cancelled. The following year, 1990, brought us one of our earliest springs on record, with temperatures over 19 °C (66 °F) in late February, and near 22 °C (72 °F) mid-March. April fooled us on the 1st with more warm sunshine and a high of 21.2 °C (70.2 °F) at Heathrow Airport – the warmest April Fools' Day since 1907 – but an invasion of Arctic air brought seven consecutive night frosts from the 3rd to the 9th, causing widespread damage to the burgeoning spring growth in orchards, on farms and smallholdings, and in ordinary suburban back gardens. On the night of the 4th/5th, the temperature fell to −9.0 °C (15.8 °F) at Grendon Underwood, near Thame, on the Bucks/Oxon border. The following night the temperature fell to −6.2 °C (20.8 °F) at the long-standing station at Rothamsted, near Harpenden in Hertfordshire – the lowest April temperature there since records began in 1873. At several places in southern and central England one or other of these two nights was the coldest of the entire year.

A larger-scale switch happened in April 1981. The first half of the month was the warmest for 20 years, but the second half of the month was the coldest since before 1900. The temperature reached 22 °C (72 °F) in southern England and the south Midlands between the 9th and 11th, and although it turned somewhat cooler mid-month it remained brilliantly sunny. In fact, many regions enjoyed almost unbroken sunshine between the 14th and 20th. Subsequently much colder northerly winds brought air directly from the Arctic for much of the remainder of the month.

The cold spell culminated in a series of devastating snowstorms that swept much of the country between the 24th and 26th. Most severely affected were the hill areas of the Pennines and Peak District, Wales and the Welsh borders, Exmoor and Dartmoor, and to a lesser extent the Cotswolds, the Marlborough and Berkshire Downs, and Salisbury Plain. The snow had a high water content and was driven by gale-force north-easterly winds – a catastrophic combination for power cables, trees, and livestock. Level snow was 2 feet thick on the fringes of the Peak District in Derbyshire, with drifts over 20 feet deep. Daytime temperatures remained close to freezing in the more hilly areas, and even in major Midlands cities like Nottingham and Birmingham afternoon values were no better than 2 or 3 °C (36-37 °F) on the three successive days, 24th, 25th, and 26th – unprecedentedly low for so late in the season. The subsequent thaw resulted in severe flooding across the east Midlands – locally the worst since the 1947 floods.

The 1981 cricket season opened on 22 April at Cambridge with a match between the University and Essex. On the third morning (the 24th) the wind-chill effect was so great that the umpires took the players off the field, judging conditions to be "unreasonable and dangerous". After lunch the players returned, stopping mid-afternoon for a drinks interval when hot coffee and soup were available.

THE EASTER HOLIDAY – TO FIX OR NOT TO FIX

Teeth chattering and goose pimples are often associated with the Easter weekend, when contrary to all common sense large swathes of the British public aim for the great outdoors. However, we are all aware that, as the first of the three spring bank holiday weekends, it sometimes has to contend with decidedly wintry weather. Indeed, over the years snow has been almost as likely to fall at Easter as at Christmas, and since 1970 white Easters have actually been more frequent than white Christmases.

The most disruptive Easter snowfalls in comparatively recent years happened in 1983, when most parts of Britain were affected to some extent. The hills of northeast England and the north Midlands had a period of snow late on Good Friday, then during the following afternoon a belt of snow spread southwards across Scotland, leaving eastern districts under several inches. Around London heavy falls caused problems that Saturday evening. By Easter morning itself, much of Essex lay under a blanket of snow four to six inches thick, while the Downs behind Dover reportedly had an eight-inch cover, and some roads in Kent were blocked. Further snow fell in the southeast during the early hours of Easter Monday, though it soon melted during the morning.

It is on occasions like this, when an early Easter brings wintry weather, that the question is asked whether the Easter holiday ought to be fixed – somewhere in the middle of April, perhaps, when in theory the weather ought to be rather more reliable. The question is not new. In 1927, during Stanley Baldwin's second term of office, Lord Desborough introduced a bill to fix Easter to the Sunday following the second Saturday in April. In other words, Easter Sunday would fall betwen 9 and 15 April every year. The bill reached the statute book in 1928, but the Act of Parliament never came into force because of a clause in the legislation that said that the various Christian Churches should first be consulted and their opinions taken into consideration. Getting on for seventy years later, the Churches are still considering it.

A close examination of the weather statistics shows that the weather is significantly better in mid-April compared with late March. Rain is just as likely, but snow less so, and temperatures average a degree or two (Celsius) higher. More important still is the fact that daylight is about an hour longer, so the evenings are lighter and the sun shines more. We should not forget, either, that our parks and gardens and countryside are more likely to look the part by the middle of April – more trees in leaf, more flowers in bloom, and so on. One other advantage, often overlooked, is that school and university terms would be much more regular from year to year.

The reason why the date of Easter varies is very complicated, and you have to go back to the very first Easter to begin to understand it all. The Crucifixion, according to the Bible, took place very shortly before the Jewish festival of Passover which in turn is linked to the spring equinox and the date of the full moon.

Eventually, in the 4th century, it was agreed amongst all the Christian Churches that Easter Day should be the Sunday following the first full moon that occurred on or immediately after the 21 March (rather than the equinox itself, which may also be on 20 or 22 March). Thus the earliest date for Easter Sunday is 22 March. At the other end of the spectrum, if the full moon and Sunday coincide on 21 March, we have to wait for the following full moon, which will be Monday 19 April, so Easter Day won't occur until 25 April. This is the latest possible date.

These extreme dates are very rare because there is a 1 in 29 chance of the full moon falling on a particular date, and a 1 in 7 chance of Sunday falling on a particular date, so the chances of a full moon and a Sunday occurring together on 21 March is roughly once every two centuries. The last time Easter was on 22 March was in 1818, and the last time it was on 25 April was in 1943. In the next couple of decades the earliest Easter will be on 23 March 2008, and the latest on 24 April 2011.

This is all very interesting, but you might be forgiven for thinking that it is hardly relevant to the tail-end of the 20th century. Our bank holiday weekend that is presently linked with Easter seems to have less and less connection with the Christian celebration of the Resurrection, so there is certainly an argument for divorcing the two without bothering with the idea of fixing the religious festival. Those who wish to mark Easter in the traditional way could continue to do so, as indeed they do at Whitsuntide. Meanwhile we could all enjoy a long weekend in the middle of April with the hope of rather better weather than is likely to have occurred at Christmas.

APRIL DROUGHT

Easters, of course, are not always wintry and depressing. There have been some stunning heatwaves, none more remarkable than that of 1949, when a temperature of 29.4 °C (85 °F) was recorded at Camden Square in London – not far from St Pancras station – on the Saturday, with values not much lower on the remaining days of the holiday weekend. April has a justified reputation for being a dry month, and droughts at this time of the year are regarded with apprehension in farming circles and in the water-supply industry, for a continued rainfall deficit during succeeding months could lead to serious water shortages during the summer. There have seven outstandingly dry Aprils in the last 230-odd years – the most recent in 1984 – with less than half an inch of rain (averaged over the whole of England and Wales). The long-term average rainfall for the month is 2.25 inches.

We all know when we have a drought: the grass in the back garden turns brown, we see pictures of dried-out reservoirs on the television, and the water people stop us using hosepipes. Scientists like to produce their own definitions. For instance, "A drought occurs when available moisture falls below the requirements of plant/animal communities, and below their ability to sustain the deficit without damage or excessive cost." What all this means is that a drought results from a prolonged shortage of rainfall, which, dare one say, is pretty obvious.

There are different types of drought according to your viewpoint. Most simply, a "meteorological drought" could be said to occur when rainfall during a specified period falls below a particular threshold. For example, in Britain this might be represented by a 50 per cent deficit over three months, or a 15 per cent shortfall over two years.

The water-supply companies want enough winter rain to ensure full reservoirs and adequate ground water. Thus a "hydrological drought" occurs when winter rainfall fails to counteract the soil-moisture deficit resulting from the previous summer. Such a drought can extend over a period of several years, given a succession of dry winters, even though the intervening summers might be rather damp.

"Agricultural drought" happens during the spring and summer – the growing season – when evaporation normally exceeds rainfall anyway. When that excess becomes very great, growing plants suffer stress thanks to restricted moisture and nutrients. This results in reduced growth, reduced yield in crops, and ultimately the plants will die.

One of the most outstanding droughts in meteorological history occurred in 1893. That year brought us an exceptional spring of heat, sunshine and lack of rain, with dry weather setting in during the first few days of March and lasting until early July. Averaged over the whole of England and Wales, March rainfall was 0.59 inch (24 per cent of the long-term average), in April it was just 0.39 inch (17 per cent), in May it was 1.80 inch (71 per cent), and in June 1.53 inch (55 per cent). The total for the four-month period was therefore 4.31 inches, compared with the normal value of 10 inches. The shortage of rain was most pronounced in southeast England, and the equivalent four-month figure for Kew Observatory was 2.78 inches, just 37 per cent of the average. At Mile End, in east London, no measurable rain was recorded between 4 March and 15 May – a period of 73 consecutive days – while Twickenham in Middlesex reported 72 successive days with no rain. These are the longest completely rainless periods in the country's weather archives.

April was arguably the most remarkable of the four months, with almost double the normal sunshine, and temperatures approaching or exceeding 27 °C (80 °F) daily from the 19th until the 25th. On 20 April the mercury touched 28.9 °C (84 °F) at Cambridge, the highest temperature ever recorded in April outside central London.

The meteorological literature of the day includes a fascinating letter from a Mr Charles Brook of Harewood Lodge, Meltham, Yorkshire. He measured the character of that exceptional season not by logging temperatures or sunshine amounts but by comparing the flowering dates of a variety of plants with earlier years. He noted daffodils were in flower 16 days earlier than usual, cherry blossom appeared 27 days early, and pear blossom burst 31 days ahead of normal.

APRIL DUST

It is during those exceptionally dry springs when our own British weather is most likely to generate duststorms and sandstorms. In most cases, they are pretty feeble efforts compared with the sandblasting jobs that are a regular feature of, say, Cairo or Kuwait, and they are usually fairly localized as well. But the idea of wind-borne dust or sand sweeping our normally green and pleasant land seems so alien that these phenomena have excited considerable interest amongst meteorologists over the years, and there are a number of well-documented examples in the literature.

All you need for a duststorm is a reasonably large source of dry sand or silt or soil, and a wind sufficiently powerful to lift and carry it. For instance, the fine sand of the Iraqi desert will be disturbed by a sustained wind speed of 15 to 20 mph, at 25 to 30 mph the air will be filled with blowing dust, and at 35 mph a raging sandstorm will blot out the sun and reduce horizontal visibility to a hundred yards or less. Every year, billions of tons of Iraqi sand are deposited in Saudi Arabia, Kuwait, Iran, and above all in the Gulf. In the British Isles we certainly have strong nough winds; what we lack, fortunately, are large quantities of fine, dry soil or sand exposed to the wind.

However, in East Anglia, the Fen district and Lincolnshire, spring is the time when the fields are still bare of crops and therefore open to the wind, humidity levels are often at their lowest of the year, and strong, nagging northeasterlies sometimes blow for days on end. If the soil has already been ploughed and prepared for sowing, or indeed if sowing has already taken place, the topsoil comprises a fine tilth and it is at this time that it is at its most vulnerable. Also, the bigger the fields, and the rarer the hedgerows, the more exposed that soil is. These are the circumstances that produce storms of sufficient regularity for them to be dubbed "Fen Blows". Minor Fen Blows occur in most years, usually between mid-March and mid-May, although they occasionally occur in other months, but a major Blow happens about once every five or six years on average.

The mechanisms that drive duststorms and sandstorms are rather different. A duststorm is caused by the wind lifting very tiny particles that are so light that they are simply carried forward by the wind with the force of gravity having very little effect until a large enough barrier to the wind causes sufficient disruption to the wind for the dust to drop in a sort of drift, rather like snow. Larger sand particles travel in a different way. The grains are not lifted to any great height; rather, they travel along curved trajectories controlled by the balance between the forces of wind and gravity, but the forward momentum of a sand particle returning to earth is sufficient for it to bounce, or to dislodge one or more other particles, which are in turn carried forward by the wind. Sand deposits are found wherever the wind is slightly disturbed, allowing the effect of gravity to overcome forward movement; drains and ditches soon clog up, and ripples form on the soil surface. Surprisingly, the typical Fen Blow appears to have characteristics more akin to a sandstorm than a duststorm.

One of the worst Fen Blows on record occurred in spring 1955. After a very dry April, a southwesterly gale set in on 4 May and lasted for the best part of two days. At the peak of the gale, the wind averaged between 35 and 45 mph, with gusts as high as

65 mph – well in excess of the generally accepted threshold for lifting dry soil, which is between 20 and 25 mph. Even at the RAF base at Mildenhall, Suffolk, at some distance from the worst affected area, horizontal visibility dropped to 300 yards for a period of four hours, and the top of the dust cloud was estimated to vary between 50 and 150 feet above the ground.

According to a study of the event by M.T. Spence, the worst-hit areas were between Cambridge and Ely, and between Ely and March, especially around the village of Manea. Drainage dykes were choked with sand, and in a potato field with its typical ridge-furrow pattern the furrows were filled with a sandy soil of much lighter colour than the ridges, emphasizing the way in which the wind was sifting particles of different sizes. People outdoors also remarked on the stinging sensation on hands and faces, confirming the presence of sand grains. However, dust was present too, and many householders found a film of dust on furniture indoors after the Blow had finished. Many fields completely lost the top two inches of soil, including recently sown seeds and small seedlings that had already emerged. Fields of barley and wheat with plants one to six inches above the soil surface were generally not eroded, although the tender plants suffered much abrasive damage.

A less severe Fen Blow occurred in May 1956, and the damage on this occasion meant that many fields remained uncultivated into the summer, so that a rare summer gale that year caused a further Fen Blow during July – a very unusual event during the month when bare soil is not normally seen. Other very serious Fen Blows happened in March 1968, May 1972 and May 1975, and an unusual one occurred in early November 1980.

Sandstorms also affect parts of the British coast from time to time – after all, where else do all the sand dunes come from? It is known that, in past centuries, large tracts of farmland and some villages have been completely buried as a result of a violent gale (or a series of gales) moving dunes from one place to another – for instance, around the fringes of Carmarthen Bay, on North Uist in the Western Isles, just north of Aberdeen near the Ythan estuary, and along the southern shore of the Moray Firth. The most famous of these is believed to have happened in October 1694, when the estate and village of Culbin, between Nairn and Forres, vanished after a violent northwesterly gale covered 10 to 12 square miles of fertile farmland to a depth of up to 50 feet. Sixteen farms were buried, as was the manor house, and the entire population including the laird had to flee their homes and livings almost empty-handed.

Dust in the atmosphere makes its presence felt in other ways too. We have become familiar in recent years with occasional falls of Saharan dust, which appear to be more frequent now than they were in the first half of the century. In hot dry weather, mini-whirlwinds have been spotted – akin to the dust devils that are a more regular component of the climate in arid regions of the world. Blowing dust and other debris also feature occasionally in our cities, especially where extensive construction work is under way. Such events were quite common in inner Birmingham during the 1960s and 1970s, when large tracts of the city were being rebuilt. And, in its most boring form, dust is always present in the atmosphere to a greater or lesser extent, in the form of haze; boring, that is, until you think of the stunning sunsets and sunrises that are a direct result of atmospheric dust-haze.

7. May

"Rough winds do shake the darling buds of May,
And summer's lease hath all too short a date . . ."
SHAKESPEARE

The bard was comparing the intended recipient of the sonnet with an English summer's day, and found the summer weather wanting. Damned with faint praise, a sensitive soul might think, given the fickleness of a typical May or June or July. Strangely, rough winds are not normally a feature of May's weather; severe gales are very much the exception to the rule, although boisterous breezes are fairly common. Had he been climatologically correct, he might have worked in something about unexpected cold snaps, late frosts, or sudden hail showers, but somehow one doubts that it would have sounded quite as good.

There is a clear-cut reason why May weather seems to be so capricious: it is at this time of the year that the prevailing westerly airflow over northwest Europe is at its weakest, vigorous Atlantic depressions are infrequent, and airstreams from all points of the compass vie one with another for occupation of our territory. Why are the Atlantic westerlies at their weakest? Well, in very simple terms, because temperature contrasts between the ocean and the bordering land masses are comparatively small during the spring, and these temperature contrasts provide the energy that drives the weather systems that habitually cross the Atlantic from west to east.

It should therefore be no surprise that easterly and north-easterly winds reach their greatest frequency during the months of April and May, and this has a startling impact on the climate of western Scotland. Easterlies are normally quite dry, although they usually pick up sufficient moisture on their journey across the North Sea to cloak the east coast of Scotland in grey cloud, mist and fog for days on end - the familiar "haar" of Aberdeen and Arbroath. But the western half of Scotland is well sheltered from these directions by the mountainous bulk of the Scottish main-land, so this is clearly the best time of the year to visit the Highlands and Islands if you prefer dry sunny weather (not everyone does, of course).

Sunny weather is certainly not guaranteed, but it is much more reliable between mid-April and mid-June than it is during the remainder of the summer. On average, the sun shines for 60 to 70 hours longer in May compared with July in the Western Isles, even though the sun is above the horizon rather longer in July. May is also 20 to 30 hours sunnier than June in north-western Britain, but most eastern, central and southern parts of the kingdom count June as their sunniest month. As one would expect, the decrease in sunshine hours after May is mirrored by an increase in rainfall. At Fort William, for instance, the normal rainfall is 3.7 inches in May, 4.9 inches in June, and 6.3 inches in July.

Some Mays have been completely dominated by northeaster-lies, and in these months the contrast between the Scottish islands on the one hand, and the normally sunny south of England on the other, can be most striking. In May 1975, for example, the sun shone for a phenomenal 313.4 hours on the Isle of Rhum and for 329.2 hours on nearby Tiree, both in the Inner Hebrides, for 306 hours on Benbecula in the Outer Hebrides, and for 300 hours at Prabost on the Isle of Skye – all roughly 50 per cent above the long-term average. At the other end of the scale, Hurley and

Shinfield, both in Berkshire, totalled just 134.2 hours, some 30 per cent below normal. Rainfall ranged from 4.25 inches at Bayham Abbey, near Tunbridge Wells, to 0.51 inch in Glen Lussa on the Kintyre peninsula. By the following month, wind patterns had changed and sunshine totals in June 1975 ranged from 148 hours at Craigdarroch in Sutherland to 361 hours on Guernsey.

May 1984, although not as sunny in the west of Scotland, provided an even more startling contrast between the south-eastern and northwestern segments of Britain – a brilliant month in the northwest but an appalling one in the southeast. The sunniest places were Point of Ayre on the northern tip of the Isle of Man, Stormont Castle in Belfast, and Bangor in County Down, all with 277 hours, while Cromer in Norfolk registered just 107 hours of bright sunshine. Rainfall varied between 5.16 inches at Romsey, Hampshire, to just 0.16 inch at Eskdalemuir, in Dumfriesshire, where it was the driest May this century.

These are extreme examples. The majority of Mays will have spells of sunny easterly winds alternating with cloudier and damper westerlies and southerlies, and cold showery northerlies. Occasionally, May will be dominated by gloomy westerly and southwesterly winds, but these are comparatively rare months. And, be warned, there is one thing the statistics cannot tell you, and that is whether or not the midges are biting!

THE MAD MAY DAY BANK HOLIDAY

Aeons ago, it seems, when James Callaghan was Prime Minister and David Gower had yet to win his first England cap, we in England (and Wales) enjoyed our first May Day bank holiday. The date was Monday 1 May 1978. . .and it poured all day. North of the border it was very different – pleasantly sunny and quite warm in places – but it wasn't a bank holiday in Scotland.

Sixteen May Day bank holidays on, we have had only one full weekend of dry, sunny weather, and we had to wait until 1990 for that. With that single exception, although there have been occasional individual days of reasonable weather, this particular bank holiday appears to have been bedevilled by cold winds or rain – or occasionally both. Those who would seek to replace it with a weekend later in the season could find plenty of ammunition in the weather record-books.

The 1979 holiday was rather better than the inaugural one – it could hardly have been worse – and southern England enjoyed a little sunshine with near-average temperatures, but for the second year running it rained heavily over Wales and parts of northern England. Spring was so late that year that the trees were only just coming into leaf even in the south. The following year the weather was dry, but a nasty, nagging nor'easter blew throughout the weekend. Although the sun shone intermittently on the Monday, it was discouragingly cold, with the temperature in some eastern counties of England failing to reach 10 °C (50 °F), and the unrelenting wind provided an enormous additional chill factor that made it feel like March rather than May.

In 1981 northern England, the Midlands and Wales were still recovering from the historic late-April snowstorm and subsequent floods. A fair Saturday was followed by a frightful Sunday of gales and heavy rain, and a cold showery Monday. Spring 1982 was generally warm but there were two short outbreaks of cold northerly winds that coincided with unerring accuracy with the Easter weekend and the May Day weekend. On the bank holiday Monday a fierce wind brought a succession of frantic showers and there were some heavy thundery downpours over the northern half of the country. In 1983 a wet weekend was very much in character for that sodden spring, and there was widespread flooding in the north Midlands and Yorkshire, while 1984 mirrored 1980 with a wicked northeast wind bringing dry but cold weather on all three days.

However, 1985 brought the first passable May Day bank holiday to London and the southeast, where sunny spells helped the temperature to 18 °C (64 °F), but there were heavy thundery showers in Wales and the West Country, while North Sea coasts remained dull and grey. In any case, most places had suffered a cool and rainy Sunday. The next year the whole of the weekend was cool and showery, which was a great pity because the Wednesday, Thursday and Friday of the preceding week had brought some wonderfully warm and sunny weather, which seemed to have broken the spell of an abysmal April. The 1987 holiday was no better, and southeast England in particular endured a weekend of frequent heavy downpours of rain and hail, many places had thunder and lightning thrown in for good measure, and daytime temperatures climbed no higher than 8 °C (46 °F) locally on the Sunday. In 1988 the holiday weekend brought that familiar mixed bag of sun and showers, but the winds were southerly, and it was warmer than average for early May, with afternoon highs mainly between 15 and 18 °C (59 and 64 °F).

In 1989, the bank holiday Monday fell on 1 May for the first time since the very first one 11 years before. The weekend weather marked a transition between one of the coldest and wettest Aprils of the century and one of the warmest and driest Mays on record. Most of Britain was cloudy but warm and humid, and rain was largely confined to Scotland and Ireland.

Which brings us to 1990. This was of course a record-breaking year. Over the country as a whole it was the hottest and sunniest year since comparable records began; a new all-time extreme temperature of 37.1 °C (98.8 °F) was logged at Cheltenham in August, and exceptional heatwaves occurred in every month except June, September, November and December. It would therefore have been grossly unfair if the May Day bank holiday had missed out yet again. The weekend fell towards the end of one of those heatwaves, and the temperature exceeded 25 °C (77 °F) daily somewhere or other in Britain from 29 April until 6 May, with a peak of 28.6 °C (83.5 °F) on 3 May at Worcester. The holiday Monday fell on 7 May, which was a few degrees cooler, but still mostly sunny – indeed it was the eleventh consecutive day of almost unbroken sunshine in the southern half of the country.

You have to have an exception to prove the rule, don't you? Just in case anyone thought that 1990 might have marked a change in regime, 1991 soon disabused them. That year's May Day holiday was one of the dullest and coldest on record, and any hardy individual who had braved the beach (such as it is) at Southend that Monday afternoon would have shivered in a temperature of just 7 °C (45° F). The clerk of the weather relented in 1992 and 1993, which were passable spring weekends with variable sunshine, moderate temperatures, and little or no rain and 1994 brought plenty of warm sunshine.

It appears that the campaign to ditch the May Day bank holiday and replace it with a Trafalgar Day holiday around 21 October has faltered, in the face of objections from other European countries. Bad as the early-May weather has been over the last 15 years, it has still outscored late October, and whatever the weather May Day has almost five hours more daylight than does Trafalgar Day.

MORE BANK-HOLIDAY GOOSE PIMPLES

It is now well over a century since Gladstone's first administration introduced the great British bank holiday. The late-spring holiday has been fixed to the last weekend in May since 1967. Before that it was linked to Whitsun, which occurs seven weeks after Easter, thus Whit Monday could in theory occur on any date between 11 May and 14 June. In practice, the earliest was 12 May 1913, while the latest was indeed on 14 June in both 1882 and 1943. Every four or five years, on average, the bank holiday weekend coincides with the Whit weekend.

As you would expect, the late-May bank holiday is rather more reliable in terms of good weather than the May Day holiday. But even late May can produce unseasonably cold weather, even with snow on occasion, and this particular holiday weekend has produced a few horrors in the past.

The worst Whit weekend occurred just over a century ago, in

1891, between 16 and 18 May. This decidedly wintry holiday followed hard on the heels of a mini-heatwave, as the meteorological observer at Hitchin (Herts) noted in his report to a contemporary journal:

> Such an instance of winter and summer in one week must be without parallel. Sunday, the 10th, snow; Wednesday, a hotter day than was ever before recorded so early in May, and snow again on the Friday, Saturday and (Whit) Sunday following.

Other correspondents reported severe frost damage to fruit and vegetables on both the Sunday and Monday mornings, notably to the potato crop, while a blanket of snow covered a large part of England, reaching a depth of 6 to 7 inches in parts of Norfolk. On Whit Sunday morning a temperature of −10.0 °C (14 °F) was logged at the Ben Nevis summit observatory – the lowest value ever authentically recorded in the UK in May. Afternoon temperatures over the English Midlands and East Anglia were around 4 or 5 °C (40-ish °F), and even in London the maximum on Whit Monday was just 6.1 °C (43 °F).

Much more recently, people living in eastern England and the Midlands may recall the leaden skies and incessant rain of the late-spring holiday in 1984. Over the Chiltern Hills it rained throughout 66 of the 72 hours, totalling nearly 2 inches, which is about the normal amount of rain for the entire month of May, and the temperature remained below 10 °C (50 °F) on all three days. Needless to say, the sun failed to put in an appearance in these regions, emulating the Whit weekend of 1889.

One of the best late-May bank holidays in recent years occurred in 1992. Large parts of the UK had practically unbroken sunshine on the Saturday, the Sunday and the Monday, and the temperature reached 26 °C (79 °F) quite widely. The equivalent weekends in both 1990 and 1989 were warm and sunny too. Never before have we had three such good ones in four years.

Arguably the hottest of these late-spring holidays occurred way back in 1933, when the long weekend fell on 3–5 June. This was in the middle of a long fine spell that brought a daily average of 14 hours' sunshine through the first eight days of June to much of eastern England. A temperature of 30 °C (86 °F) was recorded as far north as Gordon Castle near Elgin on Whit Sunday, while further south Rickmansworth in Hertfordshire logged 31.1 °C (88 °F) and Camden Square in London 31.7 °C (89 °F) on Whit Monday.

Even higher temperatures occurred on Whit Monday 1944 (which fell on 29 May) although the earlier days of that weekend were less noteworthy. Readings of 32.8 °C (91 °F) were reported from Camden Square (again) and Regent's Park, from Horsham in Sussex, and from Tunbridge Wells in Kent, equalling the highest temperature ever recorded in the United Kingdom in May. The Monday of the 1953 holiday was also very hot, with 31.7 °C (89 °F) registered at Heathrow Airport and Farnham (Surrey).

From a purely climatological point of view, the decision to fix this holiday to the last weekend in May was a bit daft. It would have made much more sense to leave it another week, until early June, when the weather is on average appreciably sunnier and warmer. Those who want to abolish the May Day bank holiday because they don't like the idea of three spring holidays in close succession might do better to turn their attention to shifting the late-spring holiday forward by a week or two.

ST PANCRAS BRINGS THE SNOW

Some exceptional weather events pass by almost unremarked, except in the darkest recesses of the official meteorological archives. One recent example was the outrageous out-of-season snowstorm that swept Scotland, Northern Ireland, and northern England on 13, 14 and 15 May 1993. You would have been hard pressed to read about it in the national press, because it simply was not mentioned. One broadsheet newspaper published an artistic picture of snow plus sheep and a shepherd, with a 56-word caption appended, but that was all. This was a splendid example of the metropolitan tunnel vision that afflicts the news media in general, including radio and TV. An inch of snow on an editor's golf course on an afternoon in February and half the next day's front page is given over to the "freak snowstorm". A foot of snow in Northumberland and Berwickshire in May, Highland and trans-Pennine routes impassable, power and telephone lines damaged, are hardly worth one jot or tittle.

That particular heavy snowfall, with a prolonged downpour of rain towards the coasts, was accurately predicted two to three days in advance, and appropriate warnings were given. It was all caused by the conjunction of gale-force northerly winds originating within a snowball's throw of the North Pole, and very warm and very humid air that had been stagnating over Britain for some days previously. The cold air, being denser, undercut the tropical air mass, forcing it upwards. This forced ascent of the warm air caused it to cool progressively, which in turn led to a prolonged condensation of the very large quantity of water vapour contained in it. Additional upward impetus was provided by the local topography.

Wet snow was observed to fall even at sea level for a time, and some slushy snow accumulated in the more hilly suburbs of Edinburgh and Glasgow. The worst conditions were above the 1000-foot contour (and that means almost one-half of Scotland and a third of northern England), with upwards of 3 inches of snow on the ground, while the higher parts of the Grampians, the Southern Uplands, the Cheviot Hills, the Lakeland fells, and also the Mourne Mountains in Ulster, had between 6 and 12 inches.

Statistics show us that snow is a fairly regular visitor to the hillier parts of Scotland during May. But snowfalls of the severity and extent of mid-May 1993 are comparatively rare. The meteorological literature contains descriptions and analyses of three previous examples that were even more extensive than this most recent storm. In May 1955 snow was observed almost daily in Scotland during the first three weeks of the month, and on 17 May snow fell for several hours across practically the whole of England and Wales, depositing an inch or so over the Cotswolds and the Chilterns, and also in the centre of Birmingham, and four inches at Malham Tarn in Yorkshire. In May 1891 substantial snowfalls occurred on the 16th, 17th, and 18th, and reportedly reached a depth of seven inches at Cossey, just outside Norwich, during the early hours of the latter date.

But arguably the most wintry weather we know of in the middle of May occurred on 17 May 1935, ruining many of the celebrations marking the Silver Jubilee of King George V. After a night of widespread damaging frosts – a minimum temperature of −8.6 °C (16.5 °F) was recorded at Rickmansworth, the infamous Hertfordshire frost hollow – snow spread southwards from Scotland, successively blanketing northern England, Wales, and finally the West Country. There were snowdrifts two feet deep in the Yorkshire Dales, a four-inch-thick cover on the normally mild Wirral peninsula, and level snow was five inches deep at Tiverton, Devon. Even the usually snow-free Isles of Scilly endured a couple of sharp snow squalls.

It may not be entirely a coincidence that each of these sudden tastes of winter occurred around the same date. Our forebears were well aware that exceptional cold snaps tended to recur in mid-May, to the extent that they became enshrined in old country weather lore. The feast day of St Pancras (the railway station and the one-time north London metropolitan borough were named after the parish church of St Pancras) is celebrated on 12 May, and he was dubbed one of the three "Ice Saints". The other two were St Servatius and St Mammertus. It was said that:

> He who shears his sheep before St Servatius' Day
> Loves his wool more than his sheep.

Enough said.

Another piece of English weather lore refers to 19 May – St Dunstan's Day. Dunstan was never regarded as one of the Ice

Saints, but he was certainly associated with sharp frosts, and the havoc that these can cause in the orchard. Dunstan was actually a long-serving Archbishop of Canterbury in the 10th century, and he is believed to have established the Benedictine Abbey at Glastonbury in Somerset, so he was clearly a pretty respectable Christian. The legend, however, is thoroughly defamatory. It tells us that Dunstan ran a brewery from Glastonbury Abbey but the business began to suffer as the cider industry grew, and at about this time cider probably overtook beer as the favoured tipple of the common people down in the West Country. Dunstan, it is alleged, sold his soul to the Devil in return for a May frost sharp enough to destroy the apple blossom each year, and thus wreck cider production. His Nibs agreed to provide the frost on or about 19 May.

The truth is that mid-May frosts entered the folklore because of the devastation caused in a few years and not through any regular occurrence. Occasional examples have of course happened this century. England's lowest authentically recorded May temperatures were −9.4 °C (15 °F) near Thetford on 11 May 1941, and −8.6 °C (16.5 °F) at Rickmansworth on the 17th in 1935.

WHEN ALL HELL BREAKS LOOSE . . .

Apart from its susceptibility to damaging frosts, May is typically a fairly peaceful month, meteorologically speaking. But from time to time all hell breaks loose, and we endure a month that is totally out of character. We have already seen that severe snowstorms can sweep large tracts of the country as late as the third week of May, providing us with unwelcome reminders of the winter just past. By contrast, some Mays bring a preview of the summer season to come, with hot humid weather leading to dramatic thunderstorms and hailstorms, with an occasional tornado thrown in for good measure.

Tornadoes, like hurricanes, hardly happen in Hertford, Hereford and Hampshire. But one certainly happened in Bedfordshire on 21 May 1950, and this particular tornado is one of the best known to have occurred in Britain. On this particular day three separate tornadoes were reported from the same part of the country, although by far the greatest proportion of damage was found along a single track, some 70 miles long, extending from Buckinghamshire across the entire breadth of Bedfordshire and Huntingdonshire, and into north Cambridgeshire.

Adjusted to 1994 prices, losses were estimated at almost three-quarters of a million pounds, with some 50 per cent of the losses incurred in the small town of Linslade (then in Bucks, now in Beds). Some fifty houses were unroofed, a bakery was demolished, farm outhouses were lifted bodily and smashed to the ground some distance away. According to local newspaper reports, many of the occupants of one migratory hen house were found to be still alive, though deeply shocked, and most of them relieved of their feathers. Another eyewitness story, also possibly apocryphal, tells of a cat in mid-air, flying past at full speed with its paws outstretched. The path of damage varied in width from 5 to 50 yards. Cars and farm vehicles were also lifted off the ground and thrown about, telephone and power lines broken, and some livestock killed. Mercifully there was no loss of human life. Other towns to report damage were Wendover and Bedford, and a bus was overturned near Ely.

British people do not regard their country as one in which tornadoes occur. This is reflected in the regularity with which mini-tornadoes and whirlwinds (which may cause no more than a few hundred pounds' worth of damage) are reported in news bulletins. Even some weather forecasters regard them as very rare events, but we now know that this is not so. An independent research body – the Tornado and Storm Research Organization, or TORRO for short – has made an exhaustive study of the phenomenon in Britain, and their efforts have revealed that tornadoes, albeit mostly minor ones, do actually occur quite frequently and sometimes in large numbers.

According to TORRO's research, there are on average 20 days

during the year when tornadoes are reported in the United Kingdom, although many of them are short-lived, typically affecting an area less than 50 yards wide by a few miles long; many do not affect populated areas and therefore cause little damage. There have been some astonishing multiple tornado outbreaks: for instance, in 1981 there was an exceptional total of 152 tornadoes occurring in just 12 days, and this included the biggest single tornado swarm in the country of 105 on 23 November that year. Localized damage was reported across an area extending from Anglesey in the west to Norfolk in the east. One notable multiple outbreak occurred on Christmas Day 1990, when a number of tornadoes were reported between Devon and Lincolnshire. Statistics show that these damaging events can occur in any month of the year and at any time of day, though they are more likely during the afternoon and early evening. Autumn is the favoured season, particularly November, although there is a secondary peak in May–June. The most frequently affected area is in eastern England from Kent and London in the south to Humberside in the north, including much of the east Midlands. Bedfordshire has the greatest reported frequency of tornadoes per unit area.

Even so, most of us have never seen a tornado "in the flesh", and our knowledge of them is largely restricted to videos, photographs and descriptions of the massively destructive tornadoes that occur quite frequently in the USA, particularly in the Midwest. This is just as well, since a major tornado is arguably the single most violent manifestation of atmospheric energy that we are likely to encounter. On 18 March 1925, a single tornado claimed the lives of 695 people as it raged across the states of Missouri, Illinois and Indiana, and others on the same day killed 97 more.

. . . THE FLOOD GATES ARE OPENED

Thunderstorms are pretty dramatic events too, although of course we are much more used to them here in Britain than we are to tornadoes. The frequency of violent storms reaches a peak at the hottest time of the year – in June, July and August – but May comes next, and over the years this month has produced some particularly vicious storms accompanied by torrential downpours.

Non-metropolitan readers no doubt smile wryly when our daily newspapers feature stories of Londoners failing to cope with out-of-the-ordinary weather. One such example occurred on the morning of Monday, 9 May 1988, when the repercussions of a night of heavy thunderstorms led to serious flooding in northwest London, which in turn resulted in traffic jams up to 20 miles long on both the M40 and the A40 – two of the main arteries running into the capital from the west. Some commuters who had set out from Oxford before seven o'clock did not reach their City offices until after midday, poor lambs.

However, it is a meteorological fact that the southeastern part of England suffers more from short intense rainstorms than any other part of the country, and a London thunderstorm is probably the nearest a stay-at-home Briton will come to a downpour of tropical intensity. On that fateful Sunday three separate storms deposited 3.5 inches of rain at Ruislip, 3.0 inches at Harrow Weald, and 2.8 inches at Northwood; in other words close on twice the average for the whole of May in a matter of hours. Indeed, at the height of the early-evening storm about an inch of rain fell in 25 minutes. It doesn't need much imagination to understand how severe flooding can hit a town or city under such circumstances, given the restricted run-off capability of an urban environment. The water-authority planners have to balance the economies of building storm drains that may only reach capacity once every 50 or 100 years against the cost of occasional but temporary dislocation.

One year later, in May 1989, no rain at all fell in some parts of the London area during the entire month. Indeed, it was one of the warmest, driest and sunniest Mays ever. But there were one or two days during the month when high temperatures triggered off localized thunderstorms, and one of these turned into a real record-breaker. The storm struck the Halifax district on the afternoon of 19 May at about three o'clock in the afternoon, and lasted little more than two hours. Eye-witnesses said that at the peak of the storm the rain fell like a solid wall of water, and was accompanied by thunder, lightning and hail.

The normal rainfall for the whole of May in Halifax is about 2.5 inches, but in the two hours of the storm the suburb of Northowram was deluged with 3.25 inches, and a phenomenal 7.6 inches was reported from Walshaw Dean reservoir high in the Pennines above Hebden Bridge. Such a large fall in such a short period is unprecedented in British records, although the infamous Hampstead storm of August 1975 deposited almost 7 inches in two and a half hours (see Chapter Ten). For this and other reasons, the Walshaw Dean figure was not accepted as an officially recognized record, although many experts regard it as "probably correct". Certainly, the flooding that resulted was devastating, and unparalleled in the district. Worst hit was the catchment area of the River Calder, especially the village of Luddenden, together with some of the outskirts of Halifax. Floodwater some four feet deep swept through some properties, while bridges and storm sewers collapsed, and cars were swept away – but there was no human loss of life.

For the residents of Halifax, flaming June that year couldn't come a moment too soon.

8. June

"Before Midsummer Day we prays for rain,
After Midsummer Day we gets it anyhow."

ANON

Climate experts call it the "June Monsoon" or the "European Monsoon". For most of the rest of us, it is simply the way that it always seems to rain when Wimbledon arrives. The very notion of cramming Henley, Ascot, Wimbledon, the Lord's Test Match, hundreds of church fêtes, and thousands of back-garden barbecues into the second half of June is an act of extreme provocation. Is it any wonder that Mother Nature cannot resist?

It is not our imagination. The tendency for the weather to change, often for the worse, during the course of June is something that was noted by our forebears long before cricket and tennis were ever thought of. As well as the old midsummer saying, noted above, ancient weather lore tells us:

When the wind shifts to the west early in June,
Expect wet weather until the end of August.

When climatologists brought scientific and statistical analysis to bear on these old adages, they found more than just a grain of truth. Britain's leading climate expert of the first half of the 20th century, C.E.P. Brooks, showed that northerly winds over northern Europe reached their highest frequency of the year around 15 June, but were rare after 20 June; meanwhile south-westerly winds blew comparatively infrequently from late March until 10 June, but were very much more common during the rest of June. These findings largely confirmed what long-term rainfall averages showed. Taking England and Wales as a whole, the driest months of the year are March, April and May, and occasionally also February and June, and these are also the months when long drags of unsettled westerly winds are least likely to occur. Averaged over the last two centuries or so, monthly rainfall is between 2.3 and 2.6 inches for each month from February to June, and between 3.2 and 3.8 inches from July onwards.

Professor Hubert Lamb, Brooks' successor as climate guru since the 1950s, appears to dislike the use of the word "monsoon", preferring the more prosaic "June Return of the Westerlies". (A "monsoon" – originating from an Arabic word meaning "season" – is the seasonal wind system in the tropics, affecting most markedly the Indian subcontinent. But to the layman the term has become intractably associated with the torrential downpours of rain that characterize the southwesterly monsoon in India. Its winter counterpart, the northeasterly monsoon, is actually a very dry wind.) Lamb has demonstrated that the "Return of the Westerlies" is recognizable in at least 70 per cent of years, and considers that it is probably the result of a coincidence of changes in the atmospheric circulation in different parts of the world that occur between mid-May and mid-June. Among the most important factors are the disappearance of snow and ice from northern Canada and Alaska, the late-spring and early-summer warmth of the northern Pacific Ocean in contrast to the northwest Atlantic Ocean, and the sudden shift of the northern hemisphere jet stream over Asia from its usual winter position south of the Himalayas to its summer location north of the Tibetan plateau. These regularly occurring links in atmospheric patterns, partly controlled by the seasons and partly by geography, have been dubbed "teleconnections".

The June Monsoon is, of course, more apparent in some years

than in others, while it is not really detectable in a quarter. variations mean that it is not sufficiently reliable to be used long-range forecasting tool. An analysis of recent Junes trates this very well. In 1993 and 1992 the westerlies arrived late – in early July in 1993 and around 30 June in 1992 – following long spells of dry warm weather in June, and a westerly type of weather then persisted for most of the rest of the summer. In 1991 and 1990 west winds set in very early – during the first week of June – after two of the driest Mays of the century, and both Junes were very cool and unsettled, but the westerlies seemed to run out of steam after about four or five weeks and warm spells followed during each July.

One very famous non-occurrence was in the record-breaking summer of 1976, when westerly winds were almost entirely absent apart from a few days in mid-June and a week or so after mid-July.

Some old country weather lore from France emphasizes the way that spells of rainy weather have a habit of beginning around mid-June; in translation they may have an ominously familiar ring to British readers:

> If it rains on St Medard's feast day
> It will rain for forty days thereafter.

and

> If St Protase brings rain
> For forty days it will remain.

St Medard's Day is 8 June, and St Protase's Day is 19 June. And, yes, St Swithin's Day falls within the forty-day limit in both cases!

SNOW IN JUNE

Those excellent Lancashire seam bowlers, Peter Lever and Peter Lee, bowled Derbyshire out twice in a day on Tuesday 3 June 1975. It was not so much a rain-affected wicket as a snow-affected one. The photograph of Buxton cricket ground under an inch or so of snow, taken the previous day, adorned several national newspapers that morning, and subsequently found its way into several almanacs, yearbooks and reference volumes. Michael Parkinson tells the (possibly apocryphal) story that the Derbyshire batsman Ashley Harvey-Walker was so cold during his second innings of 26 that he handed the umpire his false teeth to stop them chattering.

Many people recall this day's extraordinary snowfalls, surely the most outrageous thing that June has ever done to us, meteorologically speaking. The freakish nature of the cold snap was accentuated by the abrupt arrival of a heatwave just four days afterwards, heralding one of the hottest summers of the century, although that was in turn eclipsed by the historic summer of '76. However, the background to this unusual weather event is not generally remembered.

The "winter" of 1974–75 was very odd indeed. The main winter period of December to February inclusive was the warmest for over a century, and almost completely snowless. Real wintry weather was confined to the autumn and spring months, with late September snowfalls as far south as the Peak District and Snowdonia, heavy sleet in Essex and Hertfordshire around breakfast time on Monday 7 October, and widespread snowfalls around Easter 1975. on Maundy Thursday afternoon traffic ground to a halt in Birmingham city centre as six inches of snow fell in as many hours, and further snow showers fell amost daily from 26 March until 10 April. May 1975 was a dry but cold month, the coldest since 1941 in parts of southern England, and damaging frosts occurred on several nights late in the month. On the morning of 31 May the temperature fell to −3.3 °C

(26.1 °F) at Grendon Underwood, near Thame, in Buckinghamshire, −3.4° C (25.9 °F) at Easthampstead, near Bracknell, in Berkshire, −4.4 °C (24.1 °F) at Alwen, near Denbigh, in northeast Wales, and −5.1 °C (22.8 °F) at Carnwath, near Lanark. This last figure is the lowest temperature ever authentically recorded anywhere in Britain during the last four days of May.

On 1 June a new surge of arctic air swept down from polar regions; it had originated over east Greenland a few days earlier, plunged southwards across Iceland and the Faeroes, reached Scotland during the evening of the 1st, and penetrated as far as the south coast of England by midday on the 2nd. Snow settled as far south as the Clent and Lickey hills, which skirt the southern fringes of Birmingham and the Black Country, although it did not last long here, and sleet was reliably reported as far south as Portsmouth and RAF Manston, near Ramsgate. This is the latest date in the season that snow has fallen so widely in southern Britain since reliable weather recording began. The cold spell lasted two more days, and although further snowfalls were negligible, there were several sharpish night frosts. Gleneagles, between Stirling and Perth, registered −3.3 °C (26.0 °F) on 2 June, and Grendon Underwood again figured in the record books with a reading of −2.0 °C (28.4 °F) on the morning of 4 June. This was actually Derby Day, and the Clerk of the Course noted a slight hoar frost on the Epsom turf when he made his early-morning

As if the fickleness of the British weather needed any emphasizing, we were then treated to one of the most spectacular meteorological reversals of the century. The northerly wind from the Arctic blew itself out on the 4th, a transition occurred on the 5th, and a very warm southerly airstream had become established over the British Isles by the 6th with afternoon temperatures heading for 25 °C (77 °F) even in northeast Scotland. The month's officially recognized temperature happened a few days later, on the 9th, when the mercury touched 28.9 °C (84.0 °F) at Achnashellach in the Northwest Highlands.

In a bizarre tail-piece to the early-June snowfalls, the *Daily Telegraph* of 24 October 1975 published a story headlined "John Arlott given 'not out'":

> A complaint that John Arlott falsely stated in the *Guardian* that snow fell at Lord's on 2nd June and that the newspaper refused to print a correction or letter pointing out the absurdity of the report, has not been upheld by the Press Council.
>
> In a report about a county cricket match, Mr Arlott states, under a heading "Arnold Melts the Summer Snow", that "during the afternoon snow fell; hitherto unknown at Lord's in June", and later, "Howarth clearly did not trust the pitch nor relish the conditions. In fairly steady snow, he played with all his instinctive wristy aggression."
>
> The Press Council's adjudication, issued today, was: "Sleet or snow had been reported as having occurred in the London area that day. Mr Arlott says that he saw snow on his sleeve while watching play at Lord's. The statement that Howarth played 'in fairly steady snow' seems probably to have been an exaggeration and is not supported by other witnesses. But the fact that others did not themselves observe snow does not prove that Mr Arlott is deliberately lying when he says and reiterates that he did."
>
> The complaint was made by Mr S. Gomez, 66, of Bridge Lane, Golders Green.

It was a perplexing summer for John Arlott. Later, in an early heatwave, also at Lord's, he was commentating when the first Test Match streaker appeared. He was clearly not entirely familiar with the phenomenon, nor with the correct terminology, referring to the gentleman concerned as a "freaker"!

FLAMING JUNE STRIKES AGAIN

Some dates in the calendar seem to attract more than their fair share of unusual weather over the years. One of these is certainly 2 June, which is famous not only for the 1975 snow, but also for atrocious weather during the 1953 Coronation, for sharp frosts in both 1962 and 1991, and for a stunning heatwave in 1947.

Another such date is 14 June. In 1896 a sudden heatwave saw the temperature reach 31 °C (88 °F) in Cornwall and 31.7 °C (89 °F) at Colmonell in south Ayrshire – just a fraction below Scotland's all-time record. In 1903 it rained without a break in London from lunchtime on the 13th to late evening on the 15th, a total of almost 59 hours, and probably the longest period of continuous rain ever recorded in a populated area in the United Kingdom. Dramatic thunderstorms and torrential downpours of rain and hail occurred in London in 1914, in Cumbria and in the Birmingham area in 1931, over Barnsley in 1969, and across large parts of England and Wales in 1977. Meanwhile, in 1990 – Britain's sunniest year on record – no sunshine at all was recorded over a sizeable sector of eastern, southern and central England during the five consecutive days 11–15 June.

The storms of 14 June 1977 were the culmination of a five-day thundery spell and were amongst the most spectacular in the last two decades. Adjacent lightning strikes caused computer malfunctions at the Meteorological Office Headquarters at Bracknell, a man was struck and killed at St Helens, and observers in Wiltshire reported one continuous peal of thunder that lasted for over 60 seconds. Many people remarked on the vividness of the lightning flashes, sufficient during the hours of darkness to see clearly the greens, pinks and reds of the average suburban garden. The storms were accompanied by cloudbursts of tropical intensity, resulting in widespread flooding in the Midlands and southeast. At Souldrop in north Bedfordshire 2.51 inches of rain fell in little more than an hour, while at Biggin Hill in Kent 2.25 inches (about equal to the normal rainfall for the entire month of June) was recorded in just 42 minutes.

Not only were there widespread thunderstorms on 14 June 1931, but also a major tornado cut a 12-mile-long swathe through

Birmingham and its environs during that afternoon. It travelled from SSW to NNE between 1440 and 1515 GMT, and structural damage occurred along a track that varied between 200 and 800 yards wide, extending from Sparkhill in the south, through Small Heath (where the damage appeared to be worst), Bordesley Green and Erdington, to the outskirts of Sutton Coldfield in the north. Torrential downpours occurred around the same time, including an exceptional fall of 2.35 inches in 38 minutes at Cannock, and serious flooding was reported in several parts of the west Midlands.

Severe floods also affected north Lancashire and southwest Cumberland, following a tremendous storm that dumped 3.5 inches of rain in less than two hours on the small town of Ulverston, and exactly 3 inches at the gasworks in Barrow-in-Furness. Local reports suggested that much larger amounts of rain may have fallen over a small area of Waberthwaite Fell and Corney Fell near the Cumbrian village of Bootle, judging by the flood damage, but regrettably there were no rain gauges in the vicinity at the time. Bridges were demolished, metalled roads completely undermined, and millions of tons of mud and gravel deposited on farmland around the village. One man was drowned, and hundreds of sheep and cattle were lost.

The 1914 thunderstorms gained much notoriety at the time as the most violent of the storms hit London. Thatched House Lodge in Richmond Park recorded 3.7 inches of rain in 165 minutes, half of which fell in only 45 minutes. Several church spires and towers in south London were struck by lightning, as indeed were a number of trees. One such, on Wandsworth Common, was affording shelter from the downpour to four people, but it offered no protection from the lightning and they were all killed. The southwestern extremity of the District Line of the London Underground was badly affected by floodwater, and many suburban houses in the Richmond/Kingston/Wimbledon area suffered a battering from hailstones described as being "as large as small plums".

QUEEN VICTORIA'S DIAMOND JUBILEE HAILSTORM

The hail that accompanied the thunderstorm of 14 June 1914 was small beer compared with the aerial bombardment that some parts of England suffered on Midsummer Day 1897. This was in the middle of a week of celebrations held to mark Queen Victoria's Diamond Jubilee. The great royal procession in London took place on Tuesday the 22nd under mainly blue skies, warm sunshine, and afternoon temperatures of 26–27 °C (80ish on the Fahrenheit scale). But any merry making planned for Midsummer Day (24 June) was dramatically cut short by spectacular afternoon thunderstorms following a morning of great heat and humidity, temperatures reaching the 90s Fahrenheit in the London area.

The storms hit a large area from Hampshire in the south to Yorkshire in the north, but they were worst in southeast England, and in Essex (and to a lesser extent in Middlesex, Hertfordshire and Bedfordshire). The thunder and lightning were accompanied by exceptional falls of hail, which not only caused widespread damage to property and crops, but also killed livestock, and many people were quite badly injured by hailstones described as being as large as hens' eggs. These storms, together with an earlier hailstorm that had hit Seaford, Sussex, on 30 May, encouraged some grotesque politico-religious speculation in the press as to the coincidence of these destructive acts of God on the one hand and the extravagant display of British imperialism on the other.

The events of 24 June were chronicled in detail in *Symons's Meteorological Magazine*, one of the leading contemporary meteorological journals. The worst damage appeared to have happened in the Chelmsford area, especially in the parishes of Ingatestone, Margaretting and Little Baddow, but a separate storm left a trail of havoc between Luton and Biggleswade in Bedfordshire.

Extracts from the published correspondence provide a vivid picture of that extraordinary afternoon, as well as some nuggets of social history. From Danbury Rectory, Chelmsford, the Reverend J. Bridges Plumtre wrote:

> At 3 pm the storm was travelling from the due West to this hill, accompanied by unusual darkness, vivid forked lightning and a rushing wind, that came along in swirls and eddies rather than in one direct blast. At 3.04 pm rain began to fall; at 3.10 pm the wind was blowing a hurricane and huge hailstones were hurled upon us; at 3.15 pm (about) the worst was over, and at 3.30 pm the storm had passed. My rain gauge measured 0.83 inch during the 26 minutes. I should like to observe that there were no hailstones in the funnel of the gauge when I examined it, though the ground all round was strewn with them. Owing to the velocity of the wind, the hail fell at an unusually acute angle. Almost every pane of glass on the west side of the house was shivered, including a large sheet of plate glass.

Alfred Roslin of Hatfield Peverel reported:

> Fairfields, a house on my farm, had 54 windows broken out of a possible 76. The storm was preceded by a dense black cloud, making it too dark to read, and the hail and rain began at 3.05 pm, and at 3.25 pm the storm was over and the sun shining brightly. Some of the hailstones were slabs of ice larger than the palm of your hand, and Mr Corder of Writtle told me he measured one 4.5 inches by 3.5 inches which weighed 5 ounces.

Mr Coverdale of Ingatestone Hall described his experiences:

> I was driving with my son and my coachman, a man named Gray. We had just got up to Mr Kortright's house when the storm came on. We jumped out of the trap and hastened into the house. The storm descended without the slightest warning, except the blackness. First the hail and rain came down vertically, and then at a sharp angle. Down came the top of a tree at once. My man Gray – how he stood the storm I don't know – managed to get the cob under the sides of the house, and we went out to try to assist. The hailstones

> immediately riddled the umbrella which I put up, and beat me back. The pony sustained a cut right down his nose, and his whole body was covered with lumps about the size of hens' eggs. My man's chest and arms looked afterwards just as if he had had five minutes with a bruiser. They were discoloured everywhere, and full of bumps. The force with which he was struck may be estimated by the fact that although he was wearing a mackintosh, livery coat, sleeved waistcoat, and shirt, he is black and blue. His tall hat was dented in. My son got a crack on the head through his hat, and there was a bump on his head in a moment as big as an egg.

The number of birds killed was enormous, and in many districts shooting was banned during the succeeding season. Not only were young birds drowned, but old ones were struck dead by the hailstones. Crows and wood pigeons fell from the trees as if shot by guns, with their heads split open by the falling ice. Over 200 fowl were killed in the village of Little Baddow, at Margaretting Hall one lamb and several chickens were killed, and at Woolmongers Farm in Stondon 50 chickens were lost. Villagers at Langford in Bedfordshire collected a large number of dead rooks from the fields the next morning.

There was also widespread damage to property. Mile End Church in Colchester was struck by lightning, damaging the roof and clock; the clock on St Osyth Priory stopped when the tower was struck; and at Ingatestone, Harding's Farm also suffered a direct lightning strike and the chimney collapsed through the farmhouse roof. In Chelmsford, the corrugated-iron shed of Messrs Hodge and Taylor was riddled with hail "as if it had been shot at". Some 25 sheets of iron were pierced.

The heaviest rainfall reported on the day came from a separate thunderstorm in Surrey, where no hail fell: some 3.15 inches of rain was registered at West Oaks, Weybridge. However, a very rare short-period fall of 1.65 inches in just 20 minutes was measured at Luton, causing serious floods along the River Lea, including much of the town centre.

THE HEATWAVE OF THE CENTURY

Some hot spells are broken by violent thunderstorms, but some hot spells just fade gently away, and, very rarely, a few hot spells seem to go on and on for ever. We British, never satisfied by our wonderfully variegated weather, start complaining after three hot days in a row. By the end of the summer of 1976, probably the hottest for over three centuries, the nation was exhausted – probably as much by the whingeing as by the heat.

It was actually quite serious. The previous summer had also been a scorcher, and the intervening winter was very dry, so water shortages soon became a problem as the temperature soared in May and June. A Drought Bill was rushed through Parliament, water consumption was restricted as reservoirs dried out, the parched countryside turned brown throughout the land, and large tracts of heath and woodland in southern England were devastated by fires. Finally a Minister for Drought, Mr Denis Howell, was appointed to coordinate water conservation – just days before the rains finally arrived at the end of August.

The centrepiece of the summer, meteorologically speaking, was a really quite exceptional heatwave that lasted from 22 June to 16 July, during which period the temperature reached the 80s Fahrenheit daily. Even more remarkable, from 23 June to 7 July inclusive, a period of 15 consecutive days, the temperature exceeded 32 °C (90 °F) somewhere or other in the country. No previous or subsequent heatwave has produced more than four successive days of 32 °C or more in any one place. Yet in 1976 several places in southern England from Sussex to Gloucestershire experienced six days of 32 °C or more, and this was extended to seven days at Cheltenham, and to nine days at Trowbridge in Wiltshire.

Before 1976, a temperature of 35 °C (95 °F) or more had been recorded only in 1906, 1911, 1923, 1932, 1948 and 1957, and since 1976 only in 1990. But this single heatwave produced five separate days above this particular threshold, including 35.6 °C

(96.1 °F) at Southampton on 28 June, equalling the all-time June record for the UK, and 35.9 °C (96.6 °F) at Cheltenham on 3 July, the highest July figure anywhere in Britain since 1911. Further afield, there were some other interesting statistics. The maximum in Wales was 33.6 °C (92.5 °F) at Usk College of Agriculture, near Newport, Gwent, and this was the highest in the principality since 1923, although that record was smashed out of sight in August 1990. At Wauchope Forest in the Cheviots, a few miles on the Scottish side of the border, a reading of 32.4 °C (90.3 °F) on 2 July was the highest in Scotland since 1908. And a new record for Northern Ireland was established on 30 June when the mercury touched 30.8 °C (87.4 °F) at Knockarevan in County Fermanagh.

Working conditions during the hot spell were almost impossible, given that in 1976 only a minority of workplaces in Britain benefited from air-conditioning. Office workers in London were particularly hard done by, as the central area of the capital always cools down at night much less efficiently than the suburbs, so offices were already hot and stuffy when staff arrived in the morning. Public transport authorities reported a substantial number of passengers suffering from heat exhaustion on several days, and a number of people collapsed at the Wimbledon tennis championships, which coincided with the hottest weather.

Along with the high temperatures, much of the country also recorded exceptional amounts of sunshine. During the 16-day period starting on 23 June, Scole (near Diss) in Norfolk recorded 227 hours and 48 minutes' bright sunshine, an average of over 14 hours per day or 88 per cent of the theoretical maximum possible. Totals in excess of 225 hours were also noted in Sussex, Kent and Suffolk.

There were some occasional bursts of thundery activity, but broadly speaking this was one of those heatwaves that gently faded away. By 8 July the temperature failed to reach 30 °C (86 °F) anywhere in Britain, but it was not until 21 July that the highest reading in the country was below 25 °C (77 °F).

MIDSUMMER MADNESS

Schoolchildren love to point out that summer lasts only seven days. If the solstice marks the beginning of summer, usually on 21 June, and Midsummer Day is on the 24th . . . well, it's simple, isn't it? George II (1683–1760), not a native-born Englishman, was even less impressed. "The British summer," he is alleged to have declared, "consists of two fine days and a thunderstorm."

Just as we discovered that there is no sensible argument why the spring equinox should mark the first day of spring, so there is no logical reason why we should call the solstice the first day of summer. For those who keep weather records, summer begins on 1 June, but this is simply for the ease of book-keeping.

The summer solstice is the longest day, or, more precisely, the day when the sun is above the horizon longer than it is on any other date. It is therefore also the day when the sun is highest in the sky at midday. In the northern hemisphere it occurs on 21 June more frequently than on any other date, but it can also occur on 20 or 22 June. This is because it takes the earth 365 days and a bit to go round the sun, therefore the exact time of the solstice is later each year by some 5 hours and 49 minutes, until the intervention of a leap year. If the difference was exactly 6 hours – that is, a quarter of a day – that would be the end of the matter. But we have a rogue 11 minutes to get rid of, and we do that by omitting the leap day in some century years. In fact, a century year is not a leap year unless it can be divided by 400, so 1800 and 1900 were not leap years, but the year 2000 will be. The last time the solstice fell on 22 June was in 1979, and it will not again be on that date until the 22nd century. Meanwhile, the last 20 June solstice was during the 19th century, but it will next fall on that date in the year 2012.

There are even more complications caused by the fact that the earth's orbit around the sun is not symmetrical. This is why the earliest sunrise and the latest sunset are out of phase, occurring respectively on 18 June and 25 June. It is also why there is often a discrepancy between midday according to the sun, and midday according to the clock. The biggest differences occur in early November, when a sundial will register midday at 11.44 am, and in early February, when solar midday occurs at 12.14 pm.

At the summer solstice the sun is above the horizon for 16 hours 39 minutes in London, for 17 hours 35 minutes in Glasgow, and for rather more than 19 hours at the northern tip of Shetland. Southerners are often surprised to discover on their first midsummer visit north of the border that in Scotland the northern sky stays comparatively light throughout the night. In Shetland, in fine weather, it is just light enough to read at midnight for a few weeks during June and July.

Midsummer (not Midsummer's) Day has precious little to do with summer, let alone the middle of it. It is actually one of the quarter days, when in medieval times quarterly rents and similar regular payments became due. In some parts of the country, particularly in rural areas, land rents are still paid on these four days. The others are Michaelmas (29 September), Christmas Day, and Lady Day (25 March). Just for the record, the middle of the calendar year occurs on 2 July, while the middle of the weather statistician's summer is 16 July.

None of these claimants to the title of midsummer is significantly warmer than any other. In fact, the date with the highest average temperature in London is 14 July.

9. July

"In this month is St Swithin's Day,
On which if that it rain, they say,
Full forty days after it will
More or less some rain distil."

ANON

This is one of the less well-known rhymes associated with St Swithin's Day, which comes round every 15 July. In many years this date provides a typical concoction of British summer weather, some parts of the country dry all day, while others are plagued by rain or showers. Comparatively rarely is the whole of the kingdom either fine and sunny all day, or dull and wet all day. Exactly how Swithin's predictive powers are supposed to cope with such contradictory weather is not known.

The ancient weather lore associated with St Swithin's Day comes in many forms but the general thrust is always the same. Here is a better-known version of Scottish origin:

St Swithin's Day if ye do rain,
For forty days it will remain;
St Swithin's Day an' ye be fair,
For forty days 'twill rain nae mair.

Most of us are familiar with the saying, but few know the story behind it or indeed anything about the saint himself.

Swithin – or more properly Swithun – was a 9th-century monk, and probably some sort of builder or architect by training. He eventually rose to become Bishop of Winchester between the years 852 and 862, and during his tenure he was responsible for several buildings in the city, including a bridge across the River Itchen, which was one of the earliest permanent river crossings in Winchester. The city was then the capital of the kingdom of Wessex, and Swithin probably occupied an important political position, and was certainly close to the court of King Ethelwulf. Indeed, it is believed that he was for a time tutor to Alfred the Great when the latter was a boy.

Towards the end of his life he left instructions that he be buried not in the usual prominent tomb in the cathedral building, but outside, "in a vile and unworthy place under the drop of the eaves, where the sweet rain of heaven may fall upon my grave".

Initially his wishes were complied with. But his successors, so the story goes, decided that it was inappropriate for a bishop's resting place to be outside the cathedral, and they organized a huge ceremony to transfer Swithin's remains to a position inside the building. This was arranged to take place on 15 July 971. But a long dry spell broke on that day, followed by weeks of exceptionally wet weather, and the monks accepted this as an indication of divine displeasure. So Swithin's bones were left where they were.

Historical records, such as they are, do not agree with the legend. It is now generally agreed that the bishop's remains were indeed moved to a shrine within the old Saxon Cathedral Church of St Peter and St Paul (long since vanished) on or about 15 July 971, and no mention is made in contemporary chronicles of bad weather. Following the Norman Conquest, a new cathedral was built, twice the size of the old one, and after the consecration of this magnificent new church, Swithin's bones, together with the remains of all former Bishops of Winchester and the Kings and Queens of Wessex, were translated from the old building to the new one. A substantial cult flourished in the early Middle Ages around the memory of St Swithin, and during the reign of Henry VIII an attempt was made to destroy the myriad superstitions that had become associated with his name by wrecking the shrine. But of course the old sayings continued to be handed down from generation to generation.

The full version of the rhyme that heads this chapter can be found in *Poor Robin's Almanac* for 1679:

> In this month is St Swithun's Day
> On which, if that it rain, they say,
> Full forty days after which it will,
> Or more or less some rain distil.
> This Swithun was a saint, I trow,
> And Winchester's bishop also,
> Who in his time did many a feat,
> As popish legends do repeat:
> A woman having broke her eggs,
> By stumbling at another's legs,
> For which she made a woeful cry.
> St Swithun chanced for to come by,
> Who made them all as sound, or more,
> Than ever that they were before.
> But whether this were so or no,
> 'Tis more than you or I do know.
> Better it is to rise betime,
> And to make hay while sun doth shine . . .
> Than to believe in tales and lies
> Which idle monks and friars devise.

We are all aware that the St Swithin's Day saying never comes true – not literally anyway – but the original storyteller was obviously aware that summer weather patterns usually become established by the middle of July and then tend to persist until at least late August. For instance, in 1993 the weather was predominantly warm and dry in late June and early July, but a change occurred around 8/9 July, and the rest of the summer was generally unsettled and cool.

Occasionally the saying almost works. For example, in both 1989 and 1990, 15 July was sunny and very warm, and hot sunny weather predominated for most of the rest of both summers with very few days of rain. In 1989, for instance, rain fell generally in southern Britain on only 7 of the subsequent 40 days, and in 1990 on only 3.

One of the most spectacular failures was in 1976: widespread thunderstorms accompanied by torrential downpours occurred over much of England during the evening, but many places had only one day with rain between then and the end of August.

THE WORST JULY CAN DO

If you like hot weather, the last 20 years have done you proud. True, there have been some indifferent summers during that period, but there have also been 6 heatwave summers fit to rank alongside the hottest during the first three-quarters of the 20th century. Indeed, the summers of 1975 and '76, 1983 and '84, and 1989 and '90 were warmer than anything that happened between 1960 and 1974. We all (well, nearly all) believe that the summers of our childhood were better than anything since. But anyone born in, say, the late 1950s would be hard-pushed to find any prolonged spell of warm sunny weather before they reached adolescence.

Some poorer summers of recent years – 1992 and 1993 for instance – have suffered unmercifully at the hands of the press. But they have really been comparatively clement when compared with some of the rotten summers that we have had in the past. In the September 1954 issue of *Weather* the editor, clearly exasperated at newspapers' treatment of the very poor weather of that summer, was moved to comment:

> At the end of a summer whose coolness and wetness have provided subject matter for cartoonists and ideas for advertisers, it is now possible to recall that there have been such summers before ... Situated as we are at the downwind end of the Atlantic, this summer's weather is not so unlikely that atomic explosions, flying saucers or condensation trails from jet aeroplanes need be invoked in explanation.

The need to excuse and explain bad summer weather in Britain has a long though not particularly honourable history. For example, the poor summer of 1980 was quite unfairly blamed on the eruption of Mount St Helens in the northwestern USA, even though that particular volcano exploded sideways and very little volcanic dust or gas entered the upper atmosphere.

However, generally speaking, it is now thought that other large volcanic eruptions, in which gas and dust are injected into the stratosphere, only tend to depress global temperatures by less than a degree. Cool summers that may have been – at least in part – the result of this effect include those of 1816 and 1912. But the simple fact is that no summer in the last hundred years or so falls outside the normal variation of temperature either side of the long-term average.

The worst Julys are arguably those with unrelenting cool changeable weather, even though day-by-day temperature values may break no records. One of the best examples of this pattern occurred in July 1988, when the temperature failed to exceed 22 °C (72 °F) on any day in parts of the London area, or 21 °C (70 °F) in Nottingham and Bournemouth. Meanwhile at Plymouth the month's absolute highest was a mere 18.3 °C (64.9 °F), at Guernsey Airport it was 18.4 °C (65.1 °F), and at Gwennap Head near Land's End it was just 16.1 °C (61.0 °F). But the mean monthly central England temperature has been lower in 14 other Julys since 1900.

Among the coldest days ever recorded in high summer was 1 July 1980. The weather charts for the day had a decidedly autumnal look about them, with a vigorous depression crossing the UK then moving out into the North Sea, bringing a plunge of northerly winds in its wake. Eastern England was heavily overcast all day with rain on and off, and midday temperatures between 10 and 12 °C (50 to 54 °F). The wind, of course, provided a substantial chill factor.

The all-time minimum temperature for July in the United Kingdom was −3.3 °C (26.0 °F), which was recorded at the Ben Nevis summit observatory on 10 July 1888. But frosts have been known to occur locally as far south as the Thames Valley. The famous frost hollow at Rickmansworth logged −0.1 °C (31.8 °F) on 2 July 1939, while Caldecott (near Corby) reported 0.0 °C (32.0 °F) on 4 July 1965 and Newport (Shropshire) measured −0.6 °C (30.9 °F) on 31 July 1965. Another well-known frost pocket, Santon Downham near Thetford in Norfolk, recorded −1.1 °C (30.0 °F) on the morning of 1 July 1960. Even more remarkable, occurring as it did in the midst of the very hot summer of 1990, was the reading of −0.5 °C (31.1 °F) on 3 July that year at Grendon Underwood, near Thame, on the Buckinghamshire–Oxfordshire border. It has long been accepted that ground frost can cause damage to susceptible plants in particularly frost-prone areas even in midsummer, but these figures imply that on very rare occasions July ground frosts may occur widely even in southern England.

SNOW IN JULY? SURELY SOME MISTAKE . . .

It happened in 1888 according to most record books. Contemporary journals faithfully reported the accounts of summer snow, and most commentators over the years have taken these reports at face value, though there have been a few honourable exceptions. This unquestioning acceptance of old observations is odd, and does not reflect well on the judgment of climate historians over the years, because even a cursory examination of the circumstances suggests something other than snow.

There is no doubt that July 1888 was an abnormally cold month. At Kew Observatory it was the coldest July in a record stretching from 1871 to 1980, while in Professor Manley's composite record for Central England (going back to 1659) there has been no colder July since, although July 1922 was equally cold. Fractionally colder months did, however, occur in 1879, 1816, 1802 and 1695. July 1888 also occurred in the midst of a long period of consistently below-average temperatures, which raised some concern amongst meteorologists of the time that this might mark the beginning of a long-term fluctuation in the British climate.

The effects of the cold summer were interesting. Agriculture suffered enormously. Haymaking was up to five weeks late and interrupted by frequent downpours; it did not begin in some areas until early August and was not finally completed until September, and much of the hay was damaged by rain and by persistent dampness. However, there was plenty of straw, and yields of wheat and barley were reasonable, if slightly below average – dry if cool weather in late September and October was a great help in concluding the grain harvest. Root crops failed in many parts of the country, and the potato crop was devastated by disease. Fruit, as might be expected, suffered badly: few varieties of apple, pear and plum ripened properly, and flavour was described as "generally deficient" due to the marked shortage of sunshine. On the plus side, the persistently low summer temperatures were regarded as beneficial to the general health of the nation, and a certain Dr Tripe, reporting to the Royal Meteorological Society, claimed that there were 60,000 fewer deaths in 1888 compared with 1887.

Let us return to the reports of snow during that July. A summary of reports of unusual weather events can be found in the relevant issue of *British Rainfall*. These include:

> On the 7th, at Oakham (Rutland), "Snow fell for about twenty minutes in the early morning."

On the 9th, at Drumnadrochit (Inverness), "Snow on hills of 3000 feet and upwards."

On the 10th, at Mansfield (Notts), "Slight snow storm."

At Abergwesyn (Brecon), "Snow in the night."

At Cushendun (Antrim), "Two inches of snow on Nostan Mount."

Also on the 10th, "Snow fell at night on the higher mountains of Cumberland and Westmorland, and Skiddaw was white next morning. The Grampians were also white with snow."

On the 11th, at Harbledown (Canterbury), "Snowing early for twenty minutes, making the ground white."

At Cudham (west Kent), "A slight fall of snow."

At Portsmouth, "Snow."

At Northampton, "Snow in neighbourhood."

At Nottingham, "Snow at 5 am."

At Elvaston (Derby), "Very stormy; hail, sleet and snow at 5 am."

At Wetwang (East Yorks), "Snow."

At Douglas (Isle of Man), "Bitter northwest gale with snow in the neighbourhood."

On the 11th and 12th, at Welshpool, "Berwyn white with snow."

Additional reports were published in the *Meteorological Magazine* for August and September 1888. These included:

From Stroud (Glos), "Snow fell at Birdlip, between Stroud and Cheltenham on 11th."

From Barkby (Leics), "Snow on the 11th between Barkby and Leicester."

From North Shields, "Snow on the 10th."

From Loch Broom (Ullapool), "Hail storm and snow on hills on the 9th."

One further sighting is described in *London Weather*, in which J.H. Brazell wrote:

Snow was reported to have fallen at Norwood on 11 July 1888; the report stated that snow was observed on the covers of market carts

> coming into London from Norwood. The minimum temperature at Kew on the day in question was 43.6 °F (6.4 °C), and snow was also reported at Oxford and in the Isle of Wight. Therefore it seems that this isolated report of snow in London in July is probably correct.

Brazell's acceptance of the report on the flimsy evidence of coincident alleged sightings of snow from other parts of southern England is quite extraordinary, coming from a professional meteorologist. Official observations showed that minimum temperatures early on 11 July were in the range 4 to 7 °C (39 to 45 °F) over most of the country, with figures in the London area mostly near 6 or 7 °C (43–45 °F). It would be a complete physical impossibility for snow falling at those temperatures to settle even for a few seconds, and the idea that it could survive the journey on a covered cart from Norwood to London is plainly laughable. The same applies to the report of snow at Canterbury making the ground white – the minimum temperature there was 5 °C (41 °F). Snow falling sufficiently heavily to cover the ground would have reduced the air temperature (by the absorption of latent heat needed to melt the snow) to within a degree of freezing point.

Other clues can be gleaned from the synoptic weather maps compiled from official weather observations, and from the general weather pattern prevailing at the time. The charts showed a northwesterly flow covering the British Isles on the 10th, a prominent trough moving southwards across the country the following night, with the flow veering northerly during the 11th. The trough was marked by a belt of showery precipitation, and appeared typical of troughs in deeply unstable polar air which sometimes produce dramatic thundery downpours in high summer. Hail was reported widely between the 10th and the 12th, and there seems to be very little doubt that, over lowland parts of Britain, small half-melted hailstones (perhaps soft hail) had been mistaken for snow. This is something that still happens, and about once every two years we get newspaper headlines claiming "Freak Summer Snowstorms" or something similar.

The reports of snow over high ground in Scotland, and perhaps also in Wales, are a different matter. Early-morning temperatures between 4 and 7 °C (39 and 45 °F) near sea level in such a showery airstream imply a freezing level between 1500 and 3000 feet. And this is consistent with the reports of snow covering the upper slopes of the Berwyn Mountains, Skiddaw, and the Northwest Highlands. Thus lowland Britain still awaits a kosher snowfall between early June and mid-September.

AND AT THE OTHER EXTREME . . .

Enough of the ludicrous . . . July should be a month of blazing sunshine and soaring temperatures, and many of Britain's all-time-record high temperatures have occurred in this month. As a general rule, British summers are not uncomfortably warm, as they can be even in other temperate latitude countries. Washington, New York and Tokyo, for instance, frequently endure lengthy spells of energy-sapping heat and humidity, and even Paris (as the song has it) sizzles. The comparative rarity of sweltering temperatures in London until the 1970s meant that air-conditioning in the workplace was not all that common, so when the record-breaking summers of 1975 and 1976 arrived, Londoners really suffered. Since then, air-conditioned offices have become the norm.

In Britain the temperature has exceeded 32 °C (90 °F) somewhere or other in the country in just over 50 of the last 100 years. But the 35 °C (95 °F) mark has been reached in only 9 years this century (up to 1994). In 1976 the threshold was reached on five separate days, and in 1990 on three. When it comes to assessing the authenticity and comparability of these highest values of all, a variety of problems are encountered. These difficulties are perfectly illustrated by contrasting the entry for Britain's highest temperature in editions of the *Guinness Book of Records* and its sister volume, the *Guinness Book of Weather Facts and Feats*, published in the 1980s – that is, before the 1990 heatwave helped to simplify matters. The *Guinness Book of Records* used to quote 36.8 °C (98.2 °F), which occurred at Epsom, Canterbury and Raunds on 9 August 1911, while the *Guinness Book of Weather Facts and Feats* preferred a reading of 38.1 °C (100.5 °F) at Tonbridge on 22 July 1868.

Why should there be such an apparent contradiction? The trouble is that most 19th-century weather observers, although they undoubtedly carried out their observations most assiduously, did not conform to any reliable standard of exposure for their thermometers. Many were hung on north-facing walls, some even on trees, and increasingingly on specially designed stands that sheltered them from rain and direct sunlight but otherwise left them open to the elements.

Only gradually did the present-day standard, the Stevenson screen, come into general use, and indeed an open Glaisher stand remained in use alongside the Stevenson screen at the Royal Observatory, Greenwich, until 1938. The oft-quoted 37.8 °C (100 °F) on 9 August 1911 was recorded on the Glaisher stand at Greenwich; the Stevenson screen maximum on the same day was 35.9 °C (96.6 °F). In contemporary publications readings of 37.2 °C (99 °F) were noted on the same date at Isleworth in west London and Ponders End in north London, but in both cases information is lacking in respect of instruments and exposure.

The modern screen, devised in the 1860s by the well-known engineer Thomas Stevenson, is that rather familiar white box on legs, sometimes likened to a "beehive on stilts". It has louvred sides, a double roof and a slatted base, all of which allow free movement of air. But it is otherwise enclosed so that even the sun's heat reflected from the ground cannot reach the thermometers. Early versions, however, were open at the bottom. The Tonbridge record was carefully kept by a Dr G.H. Fielding, but his screen had no base and would have received heat radiating upwards from the sun-warmed soil or gravel below. There were several very hot days in the summer of 1868, and the Tonbridge readings regularly exceeded those made at nearby weather stations by a substantial amount.

When Britain's highest temperatures are talked or written about, it is nearly always the national record that is featured, and all reference books are now coming into line since the officially recognized new record of 37.1 °C (98.8 °F) was logged at Cheltenham on 3 August 1990. But this is probably of only passing interest to someone living in Scotland or Wales, while a resident of the Western Isles or Shetland would struggle to find their own local "hottest day" published anywhere.

Scotland's highest official temperature of 32.8 °C (91 °F) occurred at Dumfries on 2 July 1908, the nearest approaches in recent years being 32.4 °C (90.3 °F) at Wauchope Forest near the Northumberland border on 2 July 1976, and 32.2 °C (90.0 °F) at nearby Kelso on 2 August 1990. The northernmost instance of a maximum of 90 °F appears to be at Gordon Castle, near Elgin, on 1 September 1906, while 30.6 °C (87.0 °F) was reached as far north as Strathy on the north coast of the mainland on 11 August 1975. The record for the Western Isles is 27.2 °C (81 °F) at Benbecula on 30 July 1948, and for Orkney it is 24.4 °C (76 °F) at Kirkwall on 11 July 1911. Shetland's greatest ever heatwave seems to have occurred on 6 August 1910 when an extraordinary reading, believed authentic, of 27.8 °C (82 °F) was made at Sumburgh.

For almost 70 years the Welsh record stood at 33.9 °C (93 °F) at Newport, Gwent, on 12 July 1923. But this was beaten at no fewer than eight separate locations in the August 1990 heatwave, and the new record now stands at 35.2 °C (95.4 °F) at Hawarden Bridge – just across the border from Chester.

Similarly, the highest temperature recorded in Northern Ireland was for many years Armagh's 30.3 °C (86.6 °F), which occurred on 10 July 1934. But the memorable summer of 1976 produced a marginally higher reading on 30 June when 30.8 °C (87.4 °F) was observed at Knockarevan in the far west of County Fermanagh. A reading of 31.1 °C (88 °F) attributed to Armagh on 31 July 1943 can be found in some publications, but this was later shown to be erroneous, and the record now stands at 31.2 °C (88.2 °F) at Downpatrick on 14 July 1983.

In the Irish Republic the temperature has soared into the 90s Fahrenheit on a few occasions. Phoenix Park in Dublin recorded 33.4 °C (92.2 °F) on 16 July 1876, and 33.3 °C (92 °F) was noted at Kilkenny Castle on 26 July 1887. During the present century, 32.2 °C (90 °F) was logged at Foynes, County Limerick, on 9 July 1934, and also at Carlow on 14 July 1983. Dublin's hottest day for over a hundred years was 2 August 1990 with a temperature of 31.1 °C (88.0 °F) being recorded.

On the Isle of Man, sea breezes on hot days prevent the temperature climbing as high as in nearby mainland areas. Until recently, the island's record was 27.8 °C (82 °F) at Douglas on 30 July 1948. But 28.9 °C (84 °F) was recorded at Ronaldsway airport on 10 July 1983, and this was nearly equalled in 1990 when 28.5 °C (83.3 °F) occurred at Point of Ayre – the island's northern tip – on 2 August. As for the Channel Islands, 35.6 °C (96 °F) was registered on 19 August 1932, and 35.4 °C (95.7 °F) on 3 August 1990, both at St Helier, on Jersey.

AND THEN THE HEAVENS OPENED

High temperatures trigger off thunderstorms – sometimes. During a spell of hot and humid weather, the forecaster will often use an expression rather like this. But some of our hottest summers have been surprisingly thunder-free. Thunderstorms during the record-breaking summer of 1976 were rare events, and this was also the case in 1990 but in 1994 hot weather and thunder were regular companions.

July 1955 provided a splendid example of the contrariness of the British climate. Broadly speaking it was one of the driest and sunniest months of the 20th century, and most parts of Britain were very warm as well. On the one hand, a large swathe of

England recorded negligible rainfall and Bury St Edmunds none at all, and on the other a new all-time UK record for rain in one day was established on the 18th. On that date 11 inches of rain fell at Martinstown, near Dorchester, much of it within a ten hour period. Put another way, four months' worth of rain fell in little more than half a day, and that is an awful lot of water in a very short time. The honour of measuring this record-breaking rainfall fell to a Mr N.I. Symons, who lived at The Chantry, in the village.

The weather had been hot and thundery since 11 July, with temperatures exceeding 27 °C (81 °F) daily. Very localized downpours of tropical intensity with shortlived flooding occurred on several occasions, including a fall of 2.35 inches at Ilford on the 11th, 2.39 inches at Shaftesbury on the 13th, and 2.74 inches at Hurst Green (Sussex) on the 14th. Three days later a major crop of thunderstorms struck Kent, with 4.22 inches of rain at Boxley and 4.19 inches at Barming, both near Maidstone. But even these exceptional falls looked insignificant compared with the Dorset storm the following day.

After a warm, humid, but rather cloudy day with occasional splatterings of raindrops, it began to rain in earnest around 3.30 (BST) in the afternoon and a torrential downpour continued until roughly 8 o'clock in the evening. This first cloudburst was responsible for between 7 and 8 inches of rain in the worst hit area, with rainfall intensity peaking at about 2 inches per hour. After a lull during the evening lasting an hour or so, torrential rain returned for a couple of hours, moderating by midnight then gradually petering out during the early hours of the 19th.

Along with the fall of exactly 11 inches at Martinstown (formerly called Winterbourne St Martin), other remarkable totals included 9.50 inches at Friar Waddon, 9.0 inches at Upwey, and 8.31 inches at Elwell. The heaviest rain in major built-up areas occurred in Dorchester and Weymouth, with 7.15 inches at Westham in Weymouth, and 7.5 inches at Queen's Avenue in Dorchester. The careful analysis of the storm carried out in the Met Office after the event suggested that 12 inches or more may have fallen in the villages of Winterbourne Steepleton and Winterbourne Abbas, just upstream from Martinstown.

The area affected by the storm was comparatively small: some 14 square miles recorded 10 inches or more, while 5 inches or more was recorded over almost 200 square miles. This compared with over 700 square miles for the great Norfolk storm of August 1912. Severe flooding occurred in many places, most seriously in Weymouth itself where the marshy area known as Radipole Lake rapidly filled with floodwater, soon overflowing its borders and rushing into nearby suburban housing estates. In the steep-sided Wey valley upstream from Weymouth, true flash floods occurred with sudden waves of water rushing down steep slopes, and causing a great deal of damage to roads and walls, and carrying a huge volume of stones, gravel and soil to lower levels. The A354 Dorchester to Weymouth road was washed out in several places. In the valley of the South Winterbourne, which is where the rain was heaviest, the flooding was less severe because the underlying chalk rock absorbed a large quantity of water before releasing it again at a much reduced rate in the form of a large number of rejuvenated springs that normally only run during the winter.

On the same day a completely separate storm struck the Glamorgan town of Neath, depositing 3.59 inches of rain, of which 3.52 inches fell in two hours. At the time, this was a record for a two-hour fall in Wales. At the peak of the storm, an inch of rain fell in just 15 minutes – also a Welsh record.

Many of Britain's largest one-day falls have occurred in the West Country – indeed the nine largest falls ever recorded in England have occurred in Devon, Somerset, or Dorset. This is no accident. The potential for these tropical downpours is greatest when the air is warm and humid, and when there is a large supply of moisture upwind of the storm area. In the British Isles, these conditions are best met in southwest England during the summer and early autumn, with a sluggish flow of air from the south or southwest, probably originating over the Bay of Biscay. One of the heaviest short-period falls of rain in recent years also occurred in this region: on the morning of 9 June 1993, 4.92 inches of rain fell in nine hours at the Royal Naval Air Station at Culdrose, Cornwall, of which 2.32 inches fell in one hour, and severe floods swept through the high street in nearby Helston.

10. August

"St Bartlemy's mantle wipes dry
All the tears St Swithin can cry."

ANON

Forty days after St Swithin's Day comes the feast of St Bartholomew – on 24 August. The old saying appears to have been designed to remind our rural ancestors that, after the all-too-familiar soggy days of late July and early August, a change is bound to happen sooner or later.

As one might expect, an analysis of the statistics does not support the idea of a change in the weather around 24 August on a sufficiently regular basis for it to be identifiable over a long period of years. However, figures for London for the period 1871–1950, compiled by the Meteorological Office, show that (at least during that period) there was a marked drop in averaged daily rainfall amounts around 5 September, accompanied by a sharp rise in the frequency of completely dry days. It could be that St Bartholomew's Day was a convenient peg on which to hang this particular item of weather lore, since this was quite an important date in the Church calendar. And, of course, the eleven 'lost' days when we moved from the Julian to the Gregorian Calendar in 1752 could easily be invoked to explain most of the slippage from 24 August to 5 September.

The close link between the Church calendar and the country

calendar is clearly demonstrated by the fact that the most comprehensive assemblage of weather-related sayings – Richard Inwards' *Weather Lore* – quotes eight sayings or proverbs relating to St Bartholomew's Day. Some are contradictory, a few are rather whimsical, and others remind us that this date was widely regarded as marking the first day of autumn during earlier centuries. After all, the first hints of approaching autumn are often brought to us towards the end of August by the lengthening nights and occasional chilly starts, and the frequent early-morning mist and heavy dew.

My favourite August saying in Inwards' collection is a French one, translated as:

> August thunder promises fat grapes and fine vintages.

The meteorological rationale behind it is difficult to adduce. Maybe a thundery month is supposed to imply prevailing warmth accompanied by substantial rainfall, thus encouraging the grapes to swell, also bringing on the ripening process, and avoiding the deleterious effects of a prolonged summer drought. This does, though, seem to be a fairly shaky hypothesis, not least because August thunder in the French vineyards is frequently accompanied by dramatic and damaging hailstorms. Parts of the Dordogne and Bordeaux regions were badly hit in the Augusts of both 1993 and 1992.

In the UK the thunderiest months are from May until September, except in the far northwest of the country. In some areas, August has more thunderstorms than any other month, notably over the Pennines and in the Greater Manchester area. Early-summer thunderstorms are usually briefer than those that occur in July and August, and it appears that prolonged thundery downpours reach their maximum frequency in August. This is confirmed by the fact that August is actually the wettest month of the year over a large chunk of eastern England from Essex to Northumberland, over much of the Lancashire/Cheshire Plain, and at a few isolated spots in eastern Scotland. It is, of course, eastern

England that is furthest from the influence of the Atlantic during autumn and winter, and therefore has much lighter winter rains than regions further west.

These torrential summer rains contribute to most of the dramatic August weather events, some of which are described in detail later in this chapter. Alongside the Lynmouth disaster and the Hampstead storm, the great Norfolk rains of August 1912, the Moray floods of August 1829, and the Tweed floods of August 1948 should be highlighted. And the record books are dripping with examples of serious flood damage, violent thunderstorms and multiple lightning strikes that have happened during this particular month in the past.

By contrast, August sometimes fails to live up to its soggy reputation. We can all think of some famous August heatwaves – the record-breaking temperatures in August 1990 for instance, or the culmination of the legendary summer of 1976 – while several Augusts have passed by with hardly a drop of rain in some parts of the country. This happened in 1991 and 1983 as well as 1976. But never was it more dramatically illustrated than in August 1947, when large tracts of the UK recorded no measurable rain, including some of the usual wetspots such as Fort William and the shores of Loch Lomond.

But the final word must belong to "anon". Another ancient saying concerning St Bartholomew's Day tells us that:

> If it rains on this day
> It will rain for forty days after.

In other words, should it happen to rain on 15 July, St Swithin's 40 wet days will take us up to St Bartholomew's Day, and his 40 means yet more rain up to and including 3 October!

RAIN STOPS PROMS: THE HAMPSTEAD DELUGE

It was almost 7 o'clock on a Thursday evening. The Promenaders hurried into the Albert Hall to escape the rain, but the storm was almost over. They could see that the sky was still very black the other side of Hyde Park, somewhere over north London probably, and the lightning continued to flicker away up there as well, but the thunder was now just a distant low rumbling. It had rained quite hard for an hour or so, but no heavier than in a dozen or more summer thunder showers; in fact nearby in Kensington Gardens the rain gauge had recorded just 10.7 mm (0.42 inches) – nothing out of the ordinary.

Our concert-goers settled into their positions, trying to get as comfortable as they could in the stifling heat. The storm had cooled things off nicely outside, but in the hall it was still uncomfortably hot and humid following an afternoon high of 29 °C (84 °F). They were looking forward to a varied programme, opening with Mendelssohn's Incidental Music to *A Midsummer Night's Dream*, followed by Webern's *Passacaglia*, Opus 1, and concluding with Dvořák's Symphony No. 8. The scheduled time for the concert to start came and went, some members of the BBC Symphony Orchestra were taking up their places and began tuning up, but the Albert Hall was far from full, and there was still no sign of conductor Bernard Haitink. Something was up.

What was up was the now infamous Hampstead storm. Parts of London ground to a very watery halt following an unprecedented but very localized downpour of truly tropical intensity. The London Underground was badly disrupted, and the Metropolitan, Circle and Bakerloo lines were brought to a complete standstill as floodwaters rushed through tunnels and the electricity supply failed. British Rail services out of and into Euston, St Pancras, King's Cross and Paddington were also badly affected, while road traffic in north London was thrown into chaos. Several members of the LSO, as well as a substantial proportion of the audience, were late arriving at the Albert Hall, and the concert eventually proceeded without (at least) two orchestral players.

The date was Thursday 14 August 1975. The rain started at almost exactly 5.15 pm, British Summer Time, and ended at about 7.50 pm. The storm was centred over Hampstead Heath, and scarcely moved during the entire duration of its most intense phase. The weather station at Hampstead Observatory, on the summit of the Heath, must have been very close to the point of maximum rainfall. The rain gauge there recorded 170.8 mm (6.72 inches) of rain – something like three months' worth of rain in considerably less than three hours. A ten-minute walk away across the Heath, Golders Hill Park measured 131.3 mm (5.17 inches), while three other gauges in the vicinity – Lanchester Road in Highgate, Waterlow Park, and Parliament Hill – also logged more than 100 mm (roughly 4 inches). And yet just four miles or so away from the storm centre, in districts such as Kingsbury, Wembley, Acton, Fulham, Finsbury, the City of London, Hackney and Walthamstow, there was almost no rain at all.

Eye witness accounts, published in the *Journal of Meteorology* for October and November 1975, gave us some flavour of the immediate impact of this quite extraordinary event. Andrew Beckman observed the storm from Cricklewood:

> At 1700 (GMT), amid tremendous lightning, the rain really started. One bolt of lightning hit the ground nearby, a split second later another bolt hit the ground in almost the same place. Almost immediately came the two bangs like two guns being fired. At about 1730 the rain eased off and I heard a few hailstones falling. I then became aware of faint shadows and noticed from the other side of the house the clear sky and sunshine no more than two miles away. The hail stopped and the rain returned in full force as the lightning and thunder intensified . . . Towards 1800 the thunder started to die down and the rain turned to sheets of hail . . . there were hailstones 16 mm in diameter. The storm finally ceased around 1830 leaving

> one man drowned, two people killed by lightning, 86.4 mm (3.40 inches) of rain, piles of hail, and many areas cut off by floods.

Patricia Beckman added:

> Our road backs on to playing fields . . . At the height of the deluge, for about twenty minutes, this entire area . . . was completely covered by a heaving off-white crust of hail nearly a foot in depth, the view resembling a winter snowscape. As continuing hail tore to this ever-swelling crust was forced by the lie of the land into waves that slowly rippled down the gardens and across the playing fields like a sea of icy porridge.

From Highgate, R.M. Corfield wrote:

> The rain was heavy immediately; frequently, heavy bursts of wind and rain sent water gushing down roofs causing gutters to overflow, and visibility to become almost nil. Later, for about fifteen minutes, there was a hailstorm, with the hailstones 5 to 8 mm in diameter. The storm lasted two and a half hours, during which time my rain-gauge collected 106.7 mm (4.2 inches) of rain. The flood-water drained from the comparatively high-lying areas of Highgate and Hampstead, flooding several areas of north and northwest London. Every house in the mile-long Tufnell Park Road was flooded, at the worst to a depth of two metres, causing many people to be made homeless; church halls were opened to accommodate these people. All emergency services had to be mobilized, and . . . a week later . . . mopping up and clearing up are still in progress. Piles of saturated bedding and furniture line the street awaiting disposal by the local authority.

Donald Hatch of Mill Hill reported:

> Near Willesden Green the rain commenced at 1705 (GMT) and the most violent phase lasted until 1820. The ground was covered with marble-size hailstones. Near Finchley Road and Hendon Way hailstones were about 10 to 11 mm across . . . chipping paintwork from cars. At Branch Hill, Hampstead, hailstones were like

ping-pong balls (say 18 to 20 mm), lightning was like "machine-gun fire" and the water flowed off the heath in "waterfalls".

Now, although the point of maximum rainfall was probably within a few hundred yards of the rain gauge at Hampstead Observatory, it is highly unlikely that site of the gauge and the wettest spot coincided. Furthermore, it is probable that the Hampstead rain gauge under-recorded the actual quantity of precipitation because many of the large hailstones that fell at the peak of the storm could have bounced out of the gauge. The official observer at Hampstead, Robert Tyssen-Gee, conducted a simple experiment to verify this. He threw a number of ice-cubes almost vertically into a rain-gauge funnel, and found that approximately 30 per cent of them jumped out again, and several others broke up with some fragments remaining in the funnel and others scattering around. Thus it is by no means impossible that nearby somewhere between 175 and 200 mm (7 and 8 inches) could have fallen.

Why did it happen? A detailed investigation into the nature and cause of the storm was subsequently carried out at the Met Office. The atmospheric conditions over eastern England during that day were highly favourable for the development of severe thunder showers, and several developed over Cambridgeshire and Lincolnshire, but the only one in southeast England was the Hampstead storm. It is likely that this storm was triggered by the additional uplift to the gentle southerly airflow provided by the high ground of Hampstead and Highgate, and enhanced by the supply of particularly hot and stagnant air which had developed over inner London during the afternoon. Air movement at cloud level was negligible, so once it had developed over Hampstead, it simply stayed there, and that topographical uplift may have encouraged regeneration of the storm when it had appeared for a time to be weakening. After 1900 BST a shield of cloud, accompanied by lower temperatures and a gentle southwesterly wind, approached from the west of London; the in-flowing air therefore became cooler and slightly less humid, and the storm steadily dissipated *in situ*.

And could it happen again? Statistical calculations showed that such a large quantity of rain in under three hours probably occurs about once every 20,000 years at Hampstead, although the margin of error either side of that figure would be very large indeed. But the official study claimed that rainfall of similar intensity could occur somewhere or other in the Greater London area about once every 25 years, although the assumptions made in that calculation were distinctly dubious. For the record, similar quantities of rain in a similarly short period of time have been measured in the UK on two other occasions, both on the flanks of the Pennines. On 11 June 1956, 155 mm (6.09 inches) was reported to have fallen at Hewenden Reservoir, on the moors above Bradford, while on 19 May 1989, 193 mm (7.60 inches) was recorded at Walshaw Dean Reservoir, above Halifax (see Chapter Seven). But, as one might expect on such occasions, some doubt was thrown on the authenticity of both measurements. On the other hand, the likelihood of a rain gauge being located close to the point of maximum rainfall in that part of the country is very small indeed.

THE LYNMOUTH DISASTER

The Lynmouth flood of 15–16 August 1952 remains one of the most tragic British natural disasters in living memory. "In the darkness of a single night," as the subsequent Disaster Appeal so starkly put it, "the village was left in ruins and 34 lives were lost." Within a few hours, a picture-postcard village was turned into a scene of utter desolation, 420 people were made homeless, 23 buildings were demolished, 70 others badly damaged, and 29 bridges were swept away. Additionally, 66 motorcars were written off, over half of them having been carried away to oblivion in the Bristol Channel. The local writer S.H. Burton described the scene thus: "I knew Lynmouth well, but when I visited the town soon after, I could not find my way about. Combined bombardment from sea and air could not have wrought more fearful devastation."

This rainstorm was very different from the Hampstead storm. It was much longer-lasting, affected a much larger area, and there was very little thunder and lightning and no hail. The weather conditions preceding the disaster made an important contribution, too. But the most important factor was the local geography, which funnelled enormous quantities of water from the Exmoor plateau through Lynmouth on its very short journey to the sea.

Following a fortnight of warm sunny weather in late July, the weather changed completely during the first few days of August, and the following two weeks were very unsettled with frequent heavy falls of rain. Large tracts of the Exmoor plateau consist of peaty soils sitting on top of a stratum of impermeable ironstone, so a spell of heavy rain will rapidly saturate the peat, turning it into an extensive bog. Thus by 15 August Exmoor was thoroughly waterlogged; any further rain would immediately find its way into the gullies and combes that drain the moor, without the delay that usually follows a summer storm as the barren peaty plateau absorbs the initial downpour.

It started to rain at about 1130 BST on the 15th, and continued without a break for almost 22 hours. For much of the the rain came down in a steady downpour, but there were two periods of torrential rain lasting one to two hours each, one around 6 pm and the other at about 9 to 11 pm. In the worst-hit area the rate of rainfall may have approached 3 inches per hour during the late-evening cloudburst. Looking at the whole storm, an area of 153 square miles, encompassing practically the whole of Exmoor, was deluged with more than 4 inches of rain, some 43 square miles received 6 inches or more, and an estimated 17 square miles had over 8 inches. The highest official figure was exactly 9 inches, recorded at Longstone Barrow, which overlooks the headwaters of the West Lyn river. But an estimated 9.1 inches fell after 1945 BST at Simonstown, one of the isolated Exmoor villages, and the subsequent Met Office report suggested that the total there could have been as much as 11 to 12 inches – more than has ever been officially recorded in a 24-hour period in this country.

The East and West Lyn rivers drain roughly 40 square miles of the moor, the eastern tributary three times as much as the western one, and together they carried almost 15 million tons of water to the Bristol Channel during the storm. They both drop 1500 feet in under four miles, and they join forces just before reaching the sea in the village of Lynmouth. S.H. Burton wrote in the journal *Weather*:

> Even in times of normal rainfall the rush of their waters down the precipitous valleys is a fearful sight. In summer, it is true, in a dry spell there is no more than a trickle at the bottom of the deep combes, but rain over the Chains (the highest part of Exmoor) transforms this trickle into a torrent very quickly . . . The enormous rainfall of August 15th, the impervious Chains and the steepness of the terrain produced a flood that no barrier could resist. Boulders so huge that even bulldozers could not shift them, full-grown trees from the wooded Lyn gorge, were hurled with unbelievable violence at banks and bridges, battering them prostrate.
>
> Down on the coast, where the East and West Lyn rivers meet, Lynmouth lay helpless in the path of the water. The valley in which the town stands is very narrow and the floods did not fan out until they had . . . reached the harbour. The houses and hotels huddling under the precipitous hills and crowding close to the river bank were exposed to the full fury of the torrent and the trees and boulders.
>
> The culvert which piped the West Lyn under part of the town and into the East Lyn became choked with debris. Impatient of the least restraint, the river swung back into its old course, mowing down the houses as a child might sweep away its building-bricks when the game had become tedious.

Our knowlege of the rainfall over Exmoor during this storm owes much to the enthusiastic unpaid work of a certain Mr C.A.

Archer. Invalided out of the consular service in the 1940s, he sought an interest that would combine his love of the outdoor life with a considerable capacity for analysing figures. He chose, in cooperation with the Met Office, to establish a series of rain gauges across the highest part of Exmoor, which until then was devoid of rainfall measurement. The gauge at Longstone Barrow was in a very remote location some 1550 feet above sea level. For official purposes it was only visited once a month, but Archer, perhaps sensing something dramatic about to happen, made an additional visit on the afternoon on 14 August and made a very difficult return journey after the storm on the 16th. He observed a total for the 44-hour period between observations of 9.04 inches, but only a couple of light showers fell outside the storm, so a figure of exactly 9 inches was accepted for this location for the period of the deluge. Such dedicated amateurs still exist, and still make important contributions to our knowledge of the British climate, although sadly the value of their work is sometimes derided by one or two very average professionals.

THE FASTNET FIASCO

An August gale is a much rarer animal than a September one. But it is by no means unheard of. They happen around the coasts of England and Wales once every two or three years on average, and they regularly catch out holiday boating enthusiasts who think that August is a "safe" month and don't bother to check the coastal-waters weather forecast.

In 1992, for instance, an unusually vigorous depression tracked across Ireland and Scotland over the August bank holiday weekend, with gusts reaching 80 mph at Fair Isle (between Orkney and Shetland), and high winds and heavy rain wrecked the holiday break for practically the whole of the country. In 1989, otherwise one of the sunniest and warmest Augusts of the century, gales swept western and northern Britain on the 14th with gusts reaching 65 mph in Cornwall, Pembrokeshire and County Down; on the same day a small tornado demolished chalets at a Butlin's holiday camp in Pwllheli, Caernarvonshire. And in 1986, the bank holiday Monday was totally washed out by the remnants of Hurricane "Charley", which swept heavy driving rain across the whole of the British Isles, accompanied by high winds that gusted to 75 mph on the Devon coast; over 5 inches of rain fell locally in north Wales.

On that occasion the appalling weather was well forecast, but the violent gale that struck on 13–14 August 1979 was not. This was the gale that subsequently entered the history books as the "Fastnet Storm".

The Fastnet Race is the culmination of the Admiral's Cup series of races, organized by the Royal Ocean Racing Club, and takes place in alternate Augusts. On Saturday 11 August some 300 yachts sailed down the Solent from Cowes, including *Morning Cloud*, skippered by former Prime Minister Edward Heath, then captain of the British team. The weather was fair, the wind a light westerly, and the barometer moderately high. The small depression located on the weather charts over Nova Scotia looked quite harmless and an awfully long way away.

Forty-eight hours later, that depression had raced across the Atlantic, deepening explosively. Sustained wind speeds in the Southwest Approaches were near 50 mph, and coastal stations reported gusts near 70 mph, including one of 74 mph at Milford Haven and 85 mph at the highly exposed site at Hartland Point. These winds built up massive waves, the largest of them 40 to 45 feet high. The worst havoc occurred to the south of Ireland over an area of comparative shallows known as the Labadie Bank. These shallows, together with a veering wind, resulted in confused irregular seas with unusually short wave-periods and therefore very steep-sided waves. The navigator of one boat that crossed the Labadie Bank was reported in the *Observer* newspaper as saying: "It was like sailing through the Alps, and the wind was so strong that it was impossible to breathe while facing it." The *Observer* continued: "The steep faces of the waves and the short distances between them created such awesome conditions that strongly built 30-foot yachts were transformed into fragile toys at the mercy of the elements."

Edward Heath wrote in the *Telegraph*:

> On *Morning Cloud*, it was not until we listened to the weather forecast at ten to six on Monday evening (more than two days after we started the race), that we first heard indications that the wind would go beyond gale force later that night. This was confirmed in the shiping forecast broadcast at 15 minutes after midnight on Tuesday, August 14th, which warned us of a severe gale force 9 and locally a storm force 10. In fact, some four hours later we encountered gusts of force 11 – well over 60 knots of wind . . .
>
> *Morning Cloud* was picked up by a roaring wave and knocked down on her side about two hours after we rounded the Fastnet Rock at 2.15 last Tuesday morning. Fortunately, she righted herself and no basic damage was done to the boat or to the crew, but it made one concentrate one's mind immediately on every possible part of the boat where danger lay if the storm got stronger and the waves still greater.

According to the official report into the disaster made by the Royal Yachting Association, of the 303 starters, 85 finished, 194 retired, including *Morning Cloud*, and 23 were abandoned, of which 5 sank. A total of 136 crew members were rescued, but 15 were drowned; in addition, 4 other yachtsmen, not participating in the race, were drowned, and there was also some loss of life on land. The inquiry concluded that the blame for the catastrophe rested with the severity of the storm rather than on any failure of yacht design or inexperience. However, a number of recommendations about design were made in the final report, and criticism was made of emergency equipment and procedure and of the rules governing the carrying of radios. Surprisingly, no direct criticism of the weather forecasts was made, although there was implicit concern about the detail and frequency of the BBC shipping forecasts. It emerged later that forecasters were severely handicapped on 13 August by major computer failures.

WHAT DID YOU DO IN THE HEATWAVE, DADDY?

August 1990 opened with the most intense heatwave Britain has had since reliable temperature records began over a hundred years ago. The hot spell culminated on 3 August with a new UK record high temperature of 37.1 °C (98.8 °F) measured at Cheltenham.

Most people wilted. I would have liked to wilt, too. But the lot of a weather expert when the weather is in record-breaking mode precludes any sort of relaxation. In fact Friday, 3 August 1990 was the second-busiest day of my life, just behind 16 October 1987 – the day after Michael Fish's hurricane.

The alarm clock intruded at 4 o'clock in the morning – as usual. The temptation to press the snooze-button was resisted – as usual. The alarm was so harsh and unforgiving, I could only cope with one dose of it per morning. In any case it was still so warm; the temperature outside stood at 21 °C (70 °F) following the preceding afternoon's reading of 33 °C (91 °F). Having collated all the early-morning data and drawn up a few charts, it was clear that the day ahead would be even hotter than its predecessors; that meant the record was in danger of being broken, and that in turn would mean a day of endless interviews and incessant telephone calls lay ahead.

At a quarter to five I jumped in the little Astra for the 45-minute journey down to London and my first weather spot on LBC – then London's news-and-talk independent radio station. Dawn was just breaking as I raced down the M1, the sky was cloud-free but the eastern horizon was fringed with a thick pinky-grey haze, and some of the Hertfordshire fields were shrouded in a soft blanket of shallow mist. In London the temperature, according to the sensor attached to my car aerial, stood at 23 °C (73 °F), and it would go no lower now that the sun was rising. What maximum temperature should I go for? 35? Or 36? Surely not 37? The previous day it had reached 34 °C (93 °F) at Heathrow Airport, and as high as 36.6 °C (97.9 °F) at Worcester. But London would have a gentle breeze from the east, blowing up from the Thames estuary, so it would not be the warmest part of the country. I decided to predict a range of temperatures: ". . . 33 °C (91 °F) on the Essex and Kent shores of the Thames, 35 °C (95 °F) in central London, 36 °C (97 °F) in the western suburbs, and that will put today on a par with London's hottest day of the century so far . . ." In the event, Heathrow's maximum was 36.5 °C (97.7 °F).

Having got the first two radio spots, at 5.40 and 6.10, out of the way, I now had a chance to scan the newspapers. The heatwave had been hogging the headlines for most of the week, but on the morning of the 3rd it had been relegated to the inside pages – Iraq had just invaded Kuwait. Thanks to the *Sun*'s chief reporter, John Kay, and a carefully organized drip-feed of new statistics and of records likely to be smashed, I had been quoted on the front page of that particular organ on the Tuesday, Wednesday and Thursday, and I could hardly grumble if we had been pushed back to page four on the Friday morning. And, as I have just occasionally regaled people with since, I must be the only person who has featured on the front page of the *Sun* on three consecutive days without having had to sue!

In addition to my regular forecasts, there were special interviews on the heatwave with Andrew Neil and Michael Parkinson, inserts into news bulletins, and thorough briefings for the weather presenter – not a specialist meteorologist – on LBC Newstalk's sister station, London Talkback Radio. At 9.30 the temperature was already 28 °C (82 °F), and a quick trip down the A4 from Hammersmith took me to Sky Television's headquarters in Isleworth. They needed a couple of interviews, one short factual one for the news bulletin, and then a more discursive one to use as part of the weather round-up feature. The trouble was they wanted to do it outside in the sunshine, and after ten minutes with the sun glaring into my face, my eyes started to water. We did the longer interview three times between eye-wipes, and they cut the good bits together to make one decent piece.

I made it back home by 11.50, just in time for the 12.10 and 1.10 "down-the-line" forecasts for LBC, and there were contributions over the telephone also for Capital Radio and British Forces Broadcasting. There were probably others, too, but the scores of telephone calls have all merged into one long conversation with the passing of time. Occasionally I stopped briefly to mop up the perspiration and to top up the body fluid with lemonade or tea, but lunch went by the board – I wasn't hungry anyway.

My regular column in *Today* was quickly completed, but the news editor rang back:

"Can you do us a list of the 50 hottest places in Britain?"

"Fifty? No, not till after 7 pm when the maximum temperatures are read. Tell you what, I'll send you 4 pm temperatures for 50 places scattered across the country. The 4 pm reading and the maximum won't be very different . . ."

"7 pm's too late. Suppose it'll have to do. Can you do it by half past three?"

"4 pm temperatures by 3.30? OK, I'll do my best." He got them at 4.30.

The hottest part of the day was between 3 and 4 pm, and I now needed to turn my attention to discovering where the highest temperature in the country had occurred. The official maximum temperature readings would not be made for a few more hours, and at a large number of weather stations' recordings are only made once a day – at 9 or 10 am. Most of the news outlets would get their figures from the Met Office, who would only use the relatively small number of stations that reported at 7 pm. Worcester had logged 36.6 °C (97.9 °F) on 2 August, and the very keen observer there, Paul Damari, told me he had made 37.0 °C (98.6 °F) on the 3rd. Here was the early leader. With a light easterly breeze, the favoured locations would be the western fringes of London, the lower Severn valley, and perhaps also the middle Thames valley. Heathrow hadn't quite reached Worcester's 37.0, nor had RAF Benson (between Oxford and Reading), so I concentrated my interest on the southwest Midlands. An hour's worth of investigative meteorology allied to a little luck turned up Cheltenham's reading of 37.1 °C (98.8 °F) courtesy of the borough council's weather observer, Frank Ford. And as we all know, Cheltenham now holds the national record.

In the meantime, the London Weather Centre had discovered a reading of 37.0 °C (98.6 °F) from an automatic weather station at Nailstone in Leicestershire, and this figure found its way onto the press agency wires, and thence into most of the newspapers the following morning. I strongly queried the authenticity of this report as Nailstone was not a recognized weather-reporting station, figured in no official meteorological network, and was out of line with all adjacent stations, which all recorded between 33 and 35 °C (91.4 and 95 °F), but my protestations were contemptuously brushed aside. I cannot suppress a smug smile even now that Nailstone (the reading, not the village) subsequently sank without trace. I passed on the Cheltenham figure to "my" newspapers and radio stations, but, good journalist that I was, withheld it from everyone else until I was sure the wrong figure had gained wide currency. After learning on BBC TV's 9 o'clock news that Nailstone had broken the all-time record, I proceeded to alert the Weather Centre of my discovery, which was – how shall I put it? – less than gratefully received.

At 10.30 that evening, having written my commentary piece for the *Sunday Telegraph* on the dangers of accepting figures from automatic weather stations without question, I slumped in front of the television for a while to help wind down after a really rather hectic day. But there I was confronted with my own ugly mug, earnestly warning of the developing drought whose very existence the water authorities were complacently denying. The pre-recorded interview for London Weekend Television had been done on the Monday morning; it seemed such a long, long time before . . .

11. September

"September dries up ditches
Or breaks down bridges."
ANON

This old saying is believed to come from Portugal, where September marks the transition from the long summer drought to the beginning of the autumn rains. Occasionally it is clearly a summer month, remaining dry and sunny throughout, while in other years it is distinctly autumnal with repeated heavy downpours.

Much the same can be said of September in Britain. The month's rainfall, averaged between 1931 and 1960, includes 49 mm (1.93 inches) in London, 61 mm (2.40 inches) in Birmingham, 66 mm (2.60 inches) in Liverpool, and 67 mm (1.64 inches) in Aberdeen – all rather less than the equivalent figures for August. However, most western districts together with south-coast counties are somewhat wetter on average in September. This contrast illustrates two seasonal influences at work. The thundery rains that reach their greatest frequency during the warmest part of the year are now in decline, and this decline will be most apparent in eastern and central districts where summer thunderstorms are most frequent. However, Atlantic westerlies are beginning to reassert themselves, and this in turn increases the frequency and heaviness of rainfall over coasts and hills facing the west.

Because of the diminishing influence of thunderstorm rains, very dry Septembers occur rather more frequently than very dry Augusts, but the increasing role played by Atlantic airstreams means that very wet Septembers are also slightly more frequent than very wet Augusts. We can look at some figures to illustrate this. Averaged over the whole of England and Wales, rainfall for both August and September is close to 80 mm (3.15 inches). If we call a month "very dry" if it has 25 mm (1 inch) of rain or less, in the 100 years from 1895 to 1994 there have been just three very dry Augusts and seven very dry Septembers. And if we call a month "very wet" if it has 150 mm (roughly 6 inches) of rain or more, in the same 100-year period there have been three very wet Augusts and six very wet Septembers. In other words, both droughts and floods are more likely in September than in August.

Comparable rainfall records extend back over two centuries, and the driest September in all that time was in 1959. The rainfall total for the month in England and Wales was just 8.4 mm (0.33 inches), and quite large areas in East Anglia, the east Midlands, and also around Plymouth, recorded no measurable rain at all. At Earls Colne (near Colchester) and at Lowestoft there was no rain from 14 August to 9 October inclusive – a period of 57 consecutive days, the longest such period anywhere in the country since the spring of 1893. Much of the rest of England had only one day with significant rain. It came at the end of one of the best summers of the 20th century, and the dry weather had begun in earnest at the beginning of February, so it was no surprise that there was widespread public concern over water shortages. Many reservoirs were practically empty, and water authorities were already limiting water supply to large parts of England. Fortunately the rains came again in mid-October and the following winter was a wet one.

The wettest September in the last 200 years occurred in 1918, when the total for England and Wales was 189.5 mm (7.46 inches) – some 234 per cent of the long-term normal for the month. A scattering of places in eastern England, from Cambridge in the south to Wakefield in the north, received four times their normal amount of rain, while many areas reported significant rainfall on all 30 days during the month. Almost exactly 30 inches was measured over the fells above Borrowdale in the Lake District, and also on the flanks of Snowdon. The rain in September 1918 was the result of an unbroken train of vigorous Atlantic depressions travelling from west to east across the British Isles, rather than any individual torrential downpours. Thanks to a warm dry August, much of the harvest had been gathered in over the southern half of England, but further north the cereal harvest was badly disrupted, and the quality of grain was very poor. As one might expect, it was also a cold and frequently windy month with a marked shortage of sunshine.

The causes of major floods in September are very much the same as for August. Probably the worst are a consequence of prolonged heavy rain over a large area, such as the southeast floods of 1968, and also the widespread but rather less severe floods of September 1993. But serious flooding has also occurred on a number of occasions following torrential localized storms, usually accompanied by thunder and hail, such as the Glasgow floods in 1976 and the dramatic storms of 5 September 1958.

September is often associated with high winds rather than heavy rain. The first gales of the season frequently occur at this time of the year, sometimes associated with the remnants of former hurricanes that have wandered away from their normal stamping ground in tropical latitudes of the Atlantic and Caribbean. Late-September blows are usually dubbed "equinoctial gales" – not because they occur more frequently around the equinox than at other times, but simply because everybody remembers the first damaging winds of the autumn.

THE DAY SURREY SWAM FOR IT

Summer 1968 had been decidedly odd: the southeastern half of the country had been cool and dull and rainy, but Scotland, Ireland, much of Wales, and northwest England enjoyed a quiet, settled season with abundant sunshine and infrequent rain. A short warm spell at the end of August ended in some ferocious thunderstorms, including one at the Oval where spectators and players helped the ground staff remove millions of gallons of water from the outfield, thus giving England (and Derek Underwood) just enough time to beat Australia. Thereafter, autumn set in depressingly early, and cool changeable weather was well established by the time the kids went back to school.

During the second week of September there were some localized thundery outbreaks across East Anglia and the southeast, but Friday 13 September was quite a pleasant day with plenty of hazy sunshine, light winds, and afternoon temperatures near 20 °C (68 °F). There was no sign on the weather charts of anything out of the ordinary in the offing, and the forecast for the weekend suggested no more than an increase in showery activity in southern districts.

On the Saturday morning, heavy rain and accompanying thunder and lightning spread across much of southern England, gradually dying out during the afternoon. But a second belt of rain pushed northwards from the English Channel during the early hours of Sunday the 15th, and became practically stationary across southeast England until the following morning. It was this second prolonged downpour that was responsible for the bulk of the rainfall in southeastern counties, and in the worst hit areas most of the rain fell within a 24-hour period spanning the two days. A large area extending from the New Forest to the Thames estuary recorded over 3 inches of rain – substantially more than the normal for the entire month of September. A subsidiary narrow belt stretching from south Buckinghamshire, along the line of the Chilterns, to west Suffolk and south Norfolk also had over 3 inches. Substantial portions of Surrey, Kent, southeast London and south Essex recorded more than 6 inches of rain, and two rainfall stations in south Essex had almost 8 inches – approaching four months' worth of rain within 24 hours.

Highest rainfall totals were as follows:

Tilbury Sewage Works	201.4 mm	(7.93 in)
Stifford Waterworks	200.9 mm	(7.91 in)
Bromley	190.5 mm	(7.50 in)
South Godstone (Surrey)	189.8 mm	(7.47 in)
Stone Street (Kent)	176.2 mm	(6.94 in)
Hildenborough (Kent)	171.0 mm	(6.73 in)

The ensuing floods were devastating. Worst hit was the low-lying portion of northwest Surrey where the rivers Mole, Wey and Loddon meander towards the Thames. But widespread flooding was reported from large parts of Surrey, Kent, Greater London, south Essex, Hertfordshire and Suffolk, with more localized problems in Hampshire, Sussex, Buckinghamshire, Bedfordshire and Norfolk. The Monday morning newspapers, unaware of the severity of the floods – which did not reach their maximum extent until the next day – nevertheless had a field day. The *Daily Telegraph*'s main headline was "Floods Devastate South – Hundreds of Homes Evacuated" and the *Daily Mail* managed the rather more picturesque "The Great Deluge – Garden of England is a Paddyfield". The quantity of water flowing in the River Mole was roughly four times the previous peak flow, while at Park Mound in Sussex the River Arun rose from a trickle to a 17-foot-deep torrent.

By the Monday and Tuesday, floodwaters covered over 100 square miles of land and had penetrated some 25,000 houses. In the urban district of Esher alone some 8000 houses suffered flood damage, affecting almost one-third of the population of 62,000. A further 3500 houses were flooded in the neighbouring districts of Walton and Weybridge, and Woking. Shops in Guildford town centre were flooded to a depth of almost eight feet. Nine road and rail bridges in Surrey were severely damaged and had to be closed, and six of these suffered a major collapse. Meanwhile, most of the major roads heading southwards and westwards from London were impassable for the best part of a day, including major trunk routes such as the A2, A21, A23, A24, and A3. Enormous tracts of farmland were inundated and many farmers with fields of root crops suffered severe losses, although the cereal harvest had been completed some weeks earlier.

It was little comfort to those who spent the rest of the month clearing up the mess from their homes, but the disaster could have been incomparably worse. The floodwaters in the valleys of the Mole and Wey gradually seeped away into the River Thames, which itself was not carrying an exceptional amount of water. This was because the middle and upper parts of the Thames valley escaped the worst of the rain. Had there been three or four inches of rain across Berkshire and Oxfordshire as well, the Thames itself would have broken its banks all the way into central London, and the Surrey floods would have been even more extensive since the water would not have been able to get away so easily.

A-SWELTER IN SEPTEMBER

Sweltering is an activity we do not normally associate with September. On average, the temperature reaches the 80s Fahrenheit (around 27 °C or more) somewhere or other in the country about once every other year, but a spell of two or three days above the 80 mark over a large part of the country is comparatively rare – in the last quarter-century only 1991, 1982 and 1973 boasted such a late heatwave. Even then, with the sun significantly lower in the sky than it is in high summer, and its strength consequently diminished, the heat is less strength-sapping than it is in a similar spell in June or July.

A flick through the record books suggests that really hot weather in September has become less common as the century has progressed. However, there is nothing to suggest that this is a permanent change; it is probably just one of those things, a characteristic statistical quirk, and there is every reason to expect that the frequency of September heatwaves will increase again sooner or later.

Those statistics reveal that readings of 30 °C (86 °F) or more occurred in 16 years between 1895 and 1973, an average of once every five years, roughly. But there has been not a single occasion since, although 1 September 1991 came close with maxima of 29.7 °C (85.5 °F) at Heathrow Airport and 29.8 °C (85.6 °F) at Marholm, near Peterborough. Moreover, the mercury has not reached the 90s Fahrenheit (over 32 °C) in September for almost half a century, yet equally high readings occurred in 1929, 1926, 1919, 1911 and 1906, as well as in 1949.

The most startling late heatwave during the 20th century occurred between 31 August and 3 September 1906, with 32 °C (90 °F) exceeded over a substantial part of England on each of these four days. The peak of the hot weather came on 2 September in most areas: 35.6 °C (96.0F) was logged at Bawtry (near Doncaster), 35.0 °C (95.0 °F) at Barnet, Epsom and Maidenhead, 34.9 °C (94.8 °F) at Collyweston, near Stamford, and 34.3 °C (93.8 °F) at Whitby. 32 °C was also reached in Wales – at Hawarden Bridge, across the border from Chester – and in Scotland at Gordon Castle, near Elgin. Such readings are surely as remarkable as the all-time record of 37.1 °C (98.8 °F) established in August 1990.

August had been very warm right through, and the prolonged heat had some serious consequences. The death rate in most British cities doubled in early August, peaked in early September, and did not return to normal levels until November. Even more alarming was a five-fold increase in infant mortality – indeed the rise in the overall death rate was almost completely accounted for by infant deaths, and most of these were caused by epidemic diarrhoea. Such epidemics of gastric infections were recognized as occurring during hot summers during the 19th and early 20th century, and largely ceased as municipal sewage disposal networks were developed. That poor domestic cleanliness was a serious health hazard was perfectly illustrated when the mayor of a large Yorkshire town offered a premium of £1 for each child in his borough that reached the age of one year, cutting the infant mortality rate in his town to less than a quarter of the rate in neighbouring towns.

One particular hazard of 19th-century heatwaves was the souring of milk. But this was reduced markedly during the 1906 hot spell compared with similar spells in 1901 and 1900 thanks to the increasingly widespread use of pasteurization, which lengthened the life of the milk by 12 hours or so. However, milk production was reduced by as much as 30 per cent for several weeks because the prolonged drought had dried out the countryside and most pastures were grazed out during August.

The drought also resulted in heath and forest fires in many parts of the country. Farmland bordering onto railway lines was also badly affected, as sparks from steam engines passing by ignited tinder-dry vegetation, and stubble fields, heather and gorse all took fire very readily. Meanwhile, on the fringes of Colchester, the drying out of the grass-covered banks of the River Colne revealed the site of a large Roman villa, hitherto unknown.

Many factory workers were given the afternoons off on the Thursday and Friday (30 and 31 August), and again the following Monday. Meanwhile, the first Saturday of the football season, 1 September, saw several matches postponed until late afternoon to avoid playing during the hottest part of the day.

It was a very prosperous season for most of Britain's holiday resorts. The steamboat company that carried many Londoners to Margate reported an increase of 20,000 customers compared with the previous year, while the Isle of Man registered a 35,000 increase in the number of visitors. Blackpool boarding houses were filled to overflowing until the last week of September.

Temperatures have reached the 90s even later in September than during the 1906 hot spell. Another heatwave happened five years later, in 1911, when 34.6 °C (94.2 °F) was recorded at

Raunds in Northamptonshire on the 8th – the third consecutive day over 90 °F. But the record for the latest occurrence of a temperature of 90 or more belongs to September 1926 when a reading of 32.2 °C (90.0 °F) was logged at Camden Square, near St Pancras station, in north London on 19 September.

AND A-SHIVERING IN SEPTEMBER

The Septembers of 1911 and 1912 could hardly have been more different. At Kew Observatory, the mean maximum temperature during September was 21.1 °C (70.0 °F) in 1911, while the following year it was just 15.5 °C (59.9 °F). The 8th was the warmest day in both months – at Hampstead the temperature reached 33.9 °C (93 °F) on that date in 1911 and a mere 17.8 °C (64.0 °F) in 1912. Four days later Buxton in Derbyshire recorded an afternoon high of 8.3 °C (47.0 °F). Such are the contrasts that September can produce.

The abnormal cold apart, September 1912 was not a particularly unpleasant month. Until the closing days of the month, Britain's weather was controlled by anticyclones – high-pressure systems – and most days were quiet and dry with light winds and intermittent sunshine. Only from the 28th onwards did the weather become wet and windy. Frosts occurred frequently, but the one published report of snowfall was certainly erroneous. In the *Meteorological Magazine* for October, a Mr Spencer Russell, writing from Southwater near Horsham in Sussex, claimed:

> Sharp snow showers were experienced here on the evenings of September 10th and 11th. On the former date, at 7 pm, a well-defined pallium of mammato-cumulus spread over the sky from the north accompanied by a keen northerly wind. Between 7.15 and 7.20 pm, snowflakes of a considerable size fell.

A close examination of the weather charts for these two days showed that the air mass covering the British Isles at the time was not sufficiently cold for true snowflakes to penetrate to ground level, and the showers that affected Sussex were probably of hail. However, that in itself is unusual enough at this time of the year.

One of the most bizarre Septembers of the century occurred in 1919, when the extremes of summer and winter occurred within 10 days of each other. The cold snap culminated in a day of widespread sleet and snow showers on the 20th – the earliest date in the autumn for significant snowfall in lowland Britain during the last hundred years.

The summer of 1919 had been exceedingly variable with long periods of great heat alternating with extended spells of abnormal cold, and this variety continued into September. The second week of the month had been exceptionally hot with the mercury touching 32.2 °C (90 °F) at Raunds in Northamptonshire on the 11th, and 31.7 °C (89 °F) at Geldeston (near Beccles) on the same date. The next day was still hot in East Anglia and the southeast, but a very sharp drop in temperature took place elsewhere as north-to-northeasterly winds set in. At Nottingham, the maximum temperature on the 11th was 29.4 °C (85 °F), while on the 12th it was just 13.9 °C (57 °F). A further short warm spell occurred around the 17th and 18th, but the following day an active cold front swept southeastwards across the whole country introducing a stiff northerly airflow that had originated well within the Arctic Circle. Night-time temperatures sank below freezing in several parts of the country, and the day's high at Aberdeen on the 19th was a mere 7.8 °C (46 °F), compared with 19.4 °C (67 °F) the previous day.

By the morning of the 20th, snow covered the ground at low levels in Scotland and northern England, with a substantial covering over higher ground throughout Wales and also over Exmoor and Dartmoor. Sleet showers were observed at lower levels as far south as the Thames valley. The official weather observer at Sheepstor on Dartmoor reported:

> On the 20th, two inches of snow fell on Exmoor and (above 1600 feet) on Dartmoor. All round Princetown on the moor there was

> enough snow to track rabbits. I cannot find anyone who remembers a fall of snow like this in September.

The Reverend R.P. Dansey wrote from Kentchurch in Herefordshire:

> On looking out at 8 am on the 20th, I could scarcely believe my eyes on seeing the Black Mountains glistening with snow under a brilliant sun. The snow level was probably 1300 feet; it had all gone by 11 am – though probably not from the northern slopes, invisible from here. The Clee Hills round Ludlow had a covering of two to three inches, and the Ludlow hounds meeting had to be stopped and taken off to the lower ground owing to the snow balling in the horses' feet. In the early morning there was a slight trace lying right down in the Teme valley, not 300 feet above sea-level. September closed with severe frost, the thermometer on the morning of the 29th registering a minimum of –6 °C (21 °F), the lowest September reading in 33 years of observations. Walnuts were frozen on the trees and completely destroyed, being subsequently turned into a dry, black, crinkled pulp.

On the flanks of Snowdon, snow lay six inches deep down to the 1200-foot contour, and above about 2500 feet the mountainside remained snow-covered for about a week.

Nothing remotely similar has happened since, although an exceptionally early snowfall was reported from the higher parts of the Scottish Highlands (above 2000 feet) on 4 September 1925. The record-breaking summer of 1976 was followed by a notable cold snap in the second week of September in Scotland, and snow settled down to about 1600 feet on the 9th in the Grampians, causing difficult driving conditions on the A939 Cockbridge to Tomintoul road, and at the Devil's Elbow pass on the A93 Braemar to Glenshee road. Above 3000 feet in the Cairngorms, fresh snow was three feet deep and provided tolerably good skiing for over a week. According to records for the automatic weather station, then run by the Institute of Hydrology, the temperature on Cairn Gorm summit remained below freezing throughout the 9, 10 and 11 September.

In addition, the Septembers of 1952, 1974, 1975 and 1986 all produced a few sleet and snow showers at moderate levels in the Scottish hills, usually during the last few days of the month. In the last 50 years, the earliest autumn snowfall over lowland Britain appears to have been that of 6 October 1974 when heavy sleet and wet snow fell at several places in East Anglia and the southeast.

EX-HURRICANES BLOW AWAY THE COBWEBS

Many Septembers are characterized by a spell of pleasant late-summer weather during the first half of the month, giving way to much more disturbed conditions with rain and strong winds during the second half. If this changeover should happen within, say, a week of the autumnal equinox, weatherwise folk (especially farmers and mariners) will talk knowingly of "equinoctial gales".

Both the spring and the autumn equinoxes have long been associated with stormy winds, and over many centuries seafarers came to fear violent gales around British coastal waters during the latter parts of March and September. Early in the present century these general observations became perverted into a belief by some mariners that gales occurred *more* frequently at the equinoxes than at any other time of the year. The statistics certainly refute such an idea. What the records *do* show is that there is quite an abrupt increase in the frequency of high winds in the region of the British Isles during the second half of September, and a more gradual decrease in late March and early April. Gale-force winds are, in fact, most common in December and early January. For instance, around the coastline of southwest England winds of force 8 or more occur on average on one day in six in December, one day in 10 in March, and one day in 15 in September. It would, therefore, be correct to say that late September usually provides the first gale of the season, while late March quite often produces the last one.

Meteorologists suspected way back in the 19th century that

some autumn windstorms were somehow linked with tropical hurricanes that had been reported several days earlier on the American/Caribbean side of the Atlantic. But they could not be certain because there was insufficient observational data over the Atlantic Ocean to enable proper daily weather charts to be assembled that would have allowed such a storm to be tracked from day to day.

We now know that a true hurricane can only maintain its intensity when it is located over the warmest parts of the planet's oceans. It will dissipate rapidly after crossing a coastline and moving inland. And it will lose its energy more gradually if it stays over water but travels into temperate latitudes, particularly when the ocean temperature beneath the storm drops below 24 °C (75 °F). However, even when such a hurricane has "died", there frequently remains a large supply of unusually warm and very moist air concentrated into a comparatively small region of the middle atmosphere. If this warm and humid air is absorbed into the circulation of a common-or-garden Atlantic depression it may provide enough added energy to cause a dramatic intensification of that depression, and this is undoubtedly the cause of some (but by no means all) of the severe September gales that have swept northwest Europe over the years.

This is why some weather forecasts refer to "ex-hurricane A" or "the remnants of hurricane B". As we have seen, in most cases this is strictly speaking incorrect, but "the Atlantic depression containing the remnants of hurricane C" is an awful mouthful!

During September 1961 two former hurricanes – Betsy and Debbie – wandered into temperature latitudes, eventually bringing high winds into the vicinity of the British Isles.

Debbie was the more violent of the two. The former hurricane wandered rather aimlessly around the Azores region for a few days, with winds in its circulation mainly of the order of 60 to 80 mph, although a few reports on 12 September indicated sustained winds of about 110 mph. Over the next few days, Debbie weakened, but was then taken over by a small temperate-latitude disturbance, and subsequently headed towards Ireland. During the 16th, the storm raced on a northeasterly track close to the western coasts of Ireland and Scotland, with the central pressure dropping close to 950 millibars (28.05 inches of mercury) as it skirted Counties Galway and Mayo. At Malin Head, a very exposed lighthouse weather-observing site on the northern tip of Ireland, a sustained wind of 76 mph was recorded at the height of the gale, with a peak gust of 113 mph. Nor was this an isolated report, for gusts of 107 mph occurred at Shannon Airport in County Clare, and 106 mph at Ballykelly, near Londonderry. In Scotland, maximum gusts were 104 mph at Tiree in the Inner Hebrides and 99 mph at Duirinish, near Kyle of Lochalsh, in Wester Ross. On the Isle of Man 104 mph was also recorded at Snaefell summit.

The high winds caused serious disruption of electricity supply and telephone services in the Irish Republic, Northern Ireland, and parts of western Scotland. Road and rail traffic were seriously disrupted by fallen trees, and shipping was also badly affected, with two vessels running aground around the Irish coast. Damage to houses and public buildings was widespread, nowhere more so than in Londonderry, where a new school was practically demolished. Losses to the cereal crop, incompletely harvested at the time, were put at £11 million at 1994 prices. Meanwhile, roughly one quarter of all the trees in Baronscourt Forest, near Newtownstewart in County Tyrone, were either uprooted or broken off.

On 16 September 1978, an intense depression containing the remnants of 'Hurricane Flossie' swept past northern Scotland, with gusts of 104 mph at Fair Isle and 100 mph at Kirkwall Airport on Orkney. There was much disruption to air and sea transport, especially amongst the developing oilfields of the Northeast Shetland Basin in the far north of the North Sea. Most roads in the Scottish Highlands were blocked by falling trees. And in September 1990 two ex-tropical storms, code-named Gustav and Isidore, travelled into temperate latitudes, the latter causing widespread gales across the British Isles with the wind gusting to 90 mph in the Western and Northern Isles.

12. October

"Earlier on today apparently a woman rang the BBC and said she'd heard that there was a hurricane on the way. Well, if you are watching, don't worry, there isn't . . .."

MICHAEL FISH, 15 October 1987

If that clip of the BBC 1 Thursday lunchtime forecast has been played once it's been played a hundred times. Michael Fish is justifiably said to be rather cheesed off about it, because the clip is almost artificially terminated at that point, because the next line spoils the fun: ". . . but having said that, actually the weather will become very windy. . ."

Southern parts of England, London included, had experienced nothing like it in living memory. Weather historians believe that the Great October Storm was the most violent and most destructive windstorm to hit southern counties for almost three centuries. Hardy northerners probably wondered what all the fuss was about; after all, severe gales are two-a-penny around our northern and western fringes. That is, until they saw the pictures on TV.

Stronger winds have been recorded in central and northern Britain on several occasions in the last twenty years, but the exceptional wind speeds registered in London and the southeast were quite unprecedented in that particular corner of the country. Peak gusts included 115 mph at Shoreham in Sussex, 108 mph at

Langdon Bay, near Dover, and 106 mph at Ashford in Kent. There were several other reports from Hampshire, Sussex and Kent around the 100 mph mark, while across the Channel a gust of 137 mph was noted on the Brittany coast. But even more significant was the strength of the wind in heavily populated areas away from the often windy coastline. For instance, there was a gust of 94 mph on the roof of State House, High Holborn, which smashed the previous record for central London by nearly 20 mph, and many other inland districts noted winds peaking at 80 mph or more.

For comparison, the all-time UK windspeed record is 173 mph, recorded on the summit of Cairn Gorm, on 20 March 1986, while the all-time record for a low-level station is 142 mph at Fraserburgh on 13 February 1989. The highest gust during the Burns' Day Storm in January 1990 was 108 mph at the very exposed site at Aberporth in Cardiganshire, and the highest in the destructive gale of 2 January 1976 was 105 mph at RAF Wittering, near Peterborough. No wind measurements are available for the Great Storm of 1703, of course, but some 8000 people perished, mostly at sea, included roughly one-third of the Royal Navy. But Professor Hubert Lamb deduced that "the strongest surface winds may have been 90 knots (104 mph) or rather more and the gusts and squalls probably much stronger" – this would place the 1703 storm at least one order of magnitude greater than that of 1987.

Eighteen people died as a result of the 1987 storm, an estimated 15 million trees were uprooted or badly damaged, and insurance losses totalled over £2000 million (at 1994 prices). Yet now that time has mellowed the vivid memories of those who suffered its full fury, the event is remembered as much for the oft-repeated allegation that there was no warning. No one would claim for a moment that the forecasts for that Thursday night were very good, or even good enough, but this is a perfect illustration of the theory that if one says something often enough, people will believe it. In the words of the cynical old journalist's favourite saying, "Don't let the facts get in the way of a good story."

The facts are these. A severe storm with heavy rain and high winds had been predicted as far back as the preceding Sunday, as anyone who had watched the *Weather Forecast for Farmers* on BBC 1 just before settling down to their Sunday lunch would testify. But as the week went on, it looked more and more as though the storm centre would actually track across Kent and East Anglia, with the strongest winds on its eastern flank, thus restricting the destructive gales to France and the Low Countries. A slight change of direction during the Thursday evening saw the storm centre's course alter by just 80 to 100 miles, but that was sufficient to put the southeastern corner of England right in the track of the most violent winds. It is difficult to know why this particular Atlantic depression was able to intensify so explosively and to pack in such extraordinarily high winds.

Certainly there was a lot of energy available to feed the storm once it had started to develop. Most of the energy for such storms comes from the contrast between the two air masses of very different origin that oppose each other across the depression's warm and cold fronts. (Regions of air of contrasting temperature and humidity don't mix very well – rather like oil and water – and the sharp boundary that develops between such air masses is called a "front".) On the Thursday evening, such a strongly marked front lay over southeast England, so that at one point Gatwick Airport had a temperature of 17 °C (63 °F) and Heathrow Airport had just 8 °C (47 °F). As the storm centre screamed north eastwards across England, cold air from the Atlantic swept in behind it even faster than the depression itself was moving, with the result that the winds grew even stronger.

Many people asked whether this was a true hurricane, or whether we were right simply to refer to it as a storm or a severe gale. To meteorologists, and also to inhabitants of the Caribbean and the USA, a hurricane is a very special sort of revolving storm that only occurs in those tropical and subtropical parts of the

Atlantic, the Caribbean, and the eastern Pacific. The very worst Caribbean hurricanes – such as Hurricane Gilbert, which devastated Jamaica in September 1988 – can blast away at a sustained speed of 150 mph or more – and that is not even counting the gusts. The expression "hurricane force" is also a very specific term that is used at sea, and by those forecasters who make predicitions for shipping and for the offshore oil industry. To reach this force the wind has to average 73 mph or more, and this did not happen in the Great October Storm, except in very localized areas for a very short period. On this basis, "storm force" was the more accurate description.

This is why some top met-men appeared on radio and TV afterwards, and rather smugly claimed that it definitely was *not* a hurricane. Thus, unwittingly, they displayed the arrogance of the scientist who purloins a word from the English language, applies a very strict definition to it for his own purposes (which is fair enough), but then tries to tell the rest of the population that the ordinary meaning of the word is wrong (which is not fair enough). According to most standard dictionaries, a hurricane is "a severe, often destructive, storm". Exactly!

OCTOBER 1993 – A DOG'S BREAKFAST OF A MONTH

October is a pivotal month in the meteorological calendar. Sometimes it is almost an extension of summer – mellow days with hazy blue skies, golden sunshine, and gentle breezes, the only autumnal hints being the misty mornings and dark evenings. Sometimes, though, the stormy winds and lashing rains of early winter cannot wait to take control, while, more rarely, keen frosts, sleet flurries and a sharp north wind provide an unwelcome foretaste of the season to come. The more genial Octobers are all the more welcome because they delay thoughts of winter, and come the end of the month it may be only four months until the first hints of spring appear. But a cool rainy September and October and a late spring may mean an interminable seven or eight months of inclement, waterlogged weather to endure.

So, different Octobers have contrasting characters. But hardly ever can we have had an October like that of 1993, an October that encompassed practically all known varieties of weather within its 31 days. We had some fairly ordinary rain during the first week, torrential downpours and floods during the second, brilliant sunshine and hard frosts during the third, while the fourth week was really rather grey and dismal and boring. Snow fell in Scotland around mid-month, thunderstorms accompanied some of the cloudbursts, and the only missing element was a spell of unusual warmth. A dog's breakfast of a month we might have called it; "episodic" is the rather more prosaic word that climatologists use to describe this sort of weather.

Looking at the month as an entity, averaged over the whole of England and Wales, it was much colder, much wetter, and also much sunnier than the long-term normal. In the preceding 50 years, only two Octobers – those of 1992 and 1974 – were significantly colder. Over the same period there were 15 wetter Octobers, most recently in 1987, and only two sunnier Octobers, which were in 1971 and 1959. Subsequently, October 1994 was also sunnier. Scotland and Northern Ireland were also colder and sunnier than average.

A closer look at the statistics reveals some startling regional variations, especially in rainfall. Much of southern and eastern England and also eastern Scotland was twice as wet as normal, which is all the more remarkable when one remembers that no rain worth the mention fell after mid-month. Over 7 inches of rain fell across a swathe of east Suffolk, roughly three times the October average, and at Theberton (roughly halfway between Ipswich and Lowestoft) some 11.2 inches of rain, more than five times the local normal, fell during the 31-day period ending on 13 October. It was so wet at the beginning of the month in southeast England that Gatwick Airport had passed its normal October rainfall by breakfast-time on the 2nd.

In complete contrast, less than half the normal rain fell in western Scotland, west Wales, Northern Ireland, and around Morecambe Bay. RAF Macrihanish, on the Kintyre peninsula in Argyllshire, registered the lowest monthly rainfall of 1.13 inches, just 22 per cent of the local normal, making it the driest October in the district since 1946.

Only the northern and southern extremities of the country had sunshine totals even fractionally below the long-term average – that is, Scotland north of Aberdeen and Fort William, together with the Channel Islands. Top of the sunshine league was Penzance with roughly 170 hours of bright sunshine, some 50 per cent above normal, and the most in October in west Cornwall this century. An astonishing total of exactly 100 sunshine hours was notched up there during the 12-day period from the 14th to the 25th – an average of eight hours and twenty minutes per day.

The mid-month change to much colder but much sunnier weather happened in very short order. As the floodwaters slowly subsided, the fields glistened white with frost as temperatures plunged. Six consecutive frosty mornings is highly unusual in October, yet that is what I recorded from the 15th to the 20th at my own Bedfordshire weather station, in common with many other parts of the Midlands and southern England. Looking back through local records, it appears that the last time such a long sequence occurred so early in the season over such a large area was in 1919.

The severest frosts occurred in central and southern Scotland and also in Cumbria, although the west and south Midlands were almost as cold. The lowest officially accepted temperature was −9.9 °C (14.2 °F) at Carnwath on the morning of the 15th; Carnwath is located in the middle Clyde valley not far from Lanark, and some 25 miles southeast of Glasgow. A preliminary search through the records suggested that this was the lowest temperature recorded anywhere in the UK so early in the season. However, it was not an October record. That remains −11.7 °C (11.0 °F), recorded at Dalwhinnie, which is just north of the

Drumochter Pass in the Grampians. Heavy snow showers occurred around this time in the northern half of Scotland, and on the morning of the 16th Aberdeen Airport reported a blanket of snow almost two inches deep – the first mid-October snowfall there for twenty years.

Severe as these mid-month frosts were, it would be a mistake to think that the first frosts of the season came early in 1993. The statistics show that most parts of Britain experience the first serious air frost of the "winter" in October. The average date is around the 10th in places like Carlisle, Shrewsbury, Lincoln and Aberdeen; the 15th in Glasgow, Leeds, Birmingham, Manchester and Oxford; and the 25th in Cambridge, Gloucester, Reading and the outer suburbs of London. Only in Scottish glens north of the Central Lowlands and also in some valleys in mid- and east Wales is this average date in September. By contrast, the coastal fringe around the whole country is normally frost-free until well into November, while frost does not usually occur in parts of Cornwall, the Isle of Wight, the Isles of Scilly, Pembrokeshire, Anglesey and in some of the outer Scottish islands until December. Central London also rarely experiences frost before the end of November.

INDIAN SUMMERS

It is an unwelcome fact, but some autumns pass by without any period of pleasantly warm and settled weather. Fortunately, the records show that most years produce at least one warm spell between mid-September and mid-November. Even in 1992 – one of the coldest autumns of the 20th century – we had a five-day spell of warm sunshine at the tail-end of September with afternoon temperatures near 25 °C (77 °F) in eastern England. In 1993, though, there was nothing in the 70s after the first week of September, and, as we have seen, when the sunshine finally came in mid-October, it was accompanied by unseasonably low temperatures.

The expression "Indian Summer" almost certainly owes its origin to North America rather than the Indian subcontinent, although a few reference books favour the latter. The first reference to it dates from the 1700s, which fits in nicely with the accelerating European settlement of the New World. The North American Indians, at least those who lived on the eastern seaboard, used to depend on extended periods of fine, quiet, sunny weather at this time of the year in order to complete the harvest and to put together stores of food to see them through the long snowy winter season. One of the features of the climate of the colonies was the excessive heat and uncomfortable humidity that prevailed (and of course still prevails) between late June and early September. And the colonists clearly noted the striking contrast between the enervating summer heat and the sunny but pleasantly cool weather that typifies late September and October when hard physical labour could be undertaken without the risk of heatstroke. According to America's leading weather historian, David Ludlum, it is said in some areas that a true Indian Summer cannot occur until after the first serious frosts of the autumn – a period known as "Squaw Winter".

Here in Britain the expression was soon linked with those unseasonably warm spells of weather that occur from time to time during the autumn. However, real Indian Summer weather lasting weeks rather than days is comparatively uncommon.

A splendid recent example occurred in 1985. Summer that year had been depressingly cool and unsettled over the whole country, and it had been disastrously wet and sunless in Scotland and Northern Ireland, seriously reducing crop yields there. But a record-breaking spell of warmth and sunshine followed during September and early October. The temperature reached the 80s Fahrenheit (27 °C or more) quite widely on 1 October, the low 70s Fahrenheit (21–24 °C) on the next two or three days, and, although it cooled off considerably thereafter, the weather remained dry and fairly sunny for most of the rest of the month. As a result of the late heatwave, the month from mid-September to mid-October was the warmest for over 200 years, while total rainfall for the two calendar months at Lowestoft was just 0.75 inches – approximately one-sixth of the long-term average.

The good burghers of March (population 14,236) in Cambridgeshire no doubt enjoyed that particular Indian Summer along with the rest of us, but probably very few of them are aware that on 1 October their town set a new all-comers record for the UK for the highest temperature ever authentically recorded in October. On that day the mercury rose to 29.4 °C (84.9 °F), beating the previous record of 28.9 °C (84.0 °F), which was logged in London in 1921, thus giving rise to the famous quiz question: "When was the all-time October temperature record set in March?"

March's record did not occur in isolation. On the same day 28.4 °C (83.1 °F) was reported from Cranwell in Lincolnshire and from Mepal, near Ely, while the temperature reached 28.3 °C (82.9 °F) in Cambridge. As far north as Newcastle 25 °C (77 °F) was recorded, while at Elgin 21 °C (70 °F) was reached. Over a large chunk of eastern England, from Essex to the Scottish border, 1 October 1985 was the hottest day of the entire year. Common sense tells us that this is quite bizarre, and statistics point to it happening only once every 400 to 500 years on average.

In spite of the tradition of Indian Summers, many people are surprised just how hot October days can be. Apart from 1985, there were three other years during the 20th century when the temperature climbed into the 80s Fahrenheit during October.

In 1959 and 1921 autumn heatwaves followed long hot summers. October 1921 was a remarkable month: 25 °C (77 °F) was exceeded somewhere or other in Britain on eight of the first ten days with a peak of 28.9 °C (84.0 °F) at Kensington Palace on both the 5th and the 6th. After the greatest heat had passed, 24.4 °C (75.0 °F) was reported from London, Essex and Northamptonshire as late as the 18th. October 1959 began in similar fashion with 25 °C (77 °F) or more on the first seven days, and 28.3 °C (83.0 °F) was logged at Rugby, Faversham and Mickleham (Surrey) on the 3rd. The hot spell in October 1908 was a most unusual one, with some of the highest temperatures occurring in northern and western Britain. Indeed, the Scottish record of 25.6 °C (78.0 °F) was established at Gordon Castle near Elgin on the 2nd, the Welsh record of 26.7 °C (80.0 °F) at Betws-y-Coed on the 3rd, and the Irish record of 25.2 °C (77.4 °F) at Clongowes Wood (County Kildare) also on the 3rd.

Heatwaves in the middle and later parts of the month are, naturally, less common, though October 1990 – a year with more than its fair share of heatwaves – produced one such between the 11th and 15th. The highest value registered during that spell was 24.8 °C (76.6 °F), reported jointly on the 13th from Waltham Cross and Enfield, both on the northern outskirts of London, and Pen-y-Ffridd, near Bangor, northwest Wales. There were also several late-October warm spells during the late 1960s and early 1970s. The UK's highest reliably recorded temperature during the last ten days of the month was 22.8 °C (73.0 °F) at Portsmouth on the 26th in 1969, but more remarkable than this was a reading of 22.2 °C (72.0 °F) made at Cape Wrath – the northwesternmost tip of mainland Scotland – on 22 October 1965. It should also be remembered that 21 °C (70 °F) has twice been exceeded this century in November.

OCTOBER SNOWFALLS

Buried deep in an early volume of *Symons's Monthly Meteorological Magazine* alongside such gems as "Hailstorm at Rochefort" and "Some Deficiencies Respecting our Knowlege of Health Resorts", lies a brief article entitled "Early Snow". It lists and describes nine occasions between 1819 and 1880 (it is in fact the 1880 volume) when snow fell in the London area during October, and on three or four of these occasions the snow settled to some depth. This small piece of research was inspired by a

12-hour snowfall on 19th/20th October 1880, which deposited two to three inches of slush in the streets of central London and a blanket of snow over six inches deep in Kent and Surrey.

Probably the most remarkable of these 19th-century falls occurred on 29 October 1836, after which snow lay on the ground for five days. An inch fell in London, but there was considerably more that that in East Anglia. Some time after, the Reverend Jenyns wrote from east Cambridgeshire:

> The last week in October 1836 was remarkable for its severity, there being sharp frosts at night, and a heavy drifting fall of snow afterwards. Newmarket Heath was so covered with snow, it being the time of the races, that it was found necessary to sweep the course before the horses could run. A great deal of fruit was still ungathered and the apples in many places were hanging upon the trees coated with snow.

Near Bury St Edmunds the snow was reported to be nearly a foot deep. It is rare enough for snow to stick around for five days in southern Britain at any time before Christmas, let alone in October.

By comparison, 20th-century Octobers have been remarkably snow-free in lowland parts of England, and there are no documented examples of a widespread snow cover in southern counties. One of the nearest approaches must have come on the morning of 7 October 1974 when sleety snow was observed falling for almost an hour over much of Essex and Hertfordshire, and also over the Downs in Surrey and Kent. The following year virtually the whole of Belgium and the Netherlands was snow-covered on the morning of 13 October. The earliest lying snow in southern England in recent years was in November 1980, with three inches at St Helier on Jersey on the 6th, and an inch or two over the hills of Kent, Sussex, and Surrey.

Further north, snowfalls this early do occur from time to time even on low ground, but they are still comparatively rare. On 31 October 1934, following three days of exceptionally cold northerly winds, wintry showers fell widely, and snow covered the ground as far south as Shropshire, Staffordshire, Leicestershire and Lincolnshire. An inch of snow covered the ground at Shrewsbury, and at Belvoir Castle (near Grantham) it was two inches thick, while at West Linton, some 15 miles southwest of Edinburgh, it was reported to be up to eight inches deep. The Octobers of 1964, 1973, 1974, 1992 and 1993 all had at least one spell when the wind blew straight down from the Arctic, and showers of sleet and snow fell as far south as the Midlands, but snow accumulated on low ground only in northeast Scotland.

It seems to be a common belief these days that cold weather in autumn will lead relentlessly to a cold winter. There is, I suppose, a certain superficial logic to it, but it was not a belief held by our forefathers. Quite the contrary, in fact, because there are a number of examples of old country sayings that suggest that a cold October or November is a necessary precursor to a mild, open sort of winter. For instance:

> If October bring much frost and wind,
> Then January and February will be mild and kind.

Recent very cold Octobers – 1993, 1992, 1981 and 1974 – provide contradictory pointers. The winter of 1993–94 was mixed with several short cold spells, that of 1992–93 was generally mild and snow-free, but the winter of 1981–82 was dominated by two spells of exceptionally cold and snowy weather. In 1974–75, meanwhile, Britain had one of the warmest winters ever recorded.

Looking back over the last hundred years or so we find equally unconvincing evidence. The twenty coldest Octobers during that period were followed by six warm winters, ten average winters, and four cold winters. And of the ten most extreme Octobers, three were followed by warm winters, four by average, and three by cold ones.

WATER, WATER, EVERYWHERE

The water industry's year, as far as rainfall is concerned, runs from October one year till September the next. This is because in a normal year rainfall exceeds evaporation between October and March, thus allowing water to accumulate in the ground and in our reservoirs. Reality often confounds our desire to simplify and summarize our knowledge, so from year to year there is considerable variation in the date when the ground becomes saturated, and of course there are always large geographical differences in this date in any particular year.

For instance, after a prolonged summer drought and a very dry autumn in 1989, many parts of England did not reach "field capacity" (that is, the ground did not become saturated) until the second week of December. By contrast in 1992, following an unsettled summer and in particular a very wet and cloudy August, a large swathe of the country was pretty well waterlogged from the August bank holiday weekend onwards, and there was no respite thereafter thanks to well above average rainfall during September and November.

If field capacity has been reached before the month begins, heavy October downpours will quickly bring widespread flooding as wide areas of eastern and southern England discovered in October 1993. Even if we escape flooding to our own houses and gardens, large numbers of us still have to cope with floodwaters disrupting road and rail transport. Scientists always seem to feel the need to define the obvious, and hydrologists have come up with some hilarious examples in an attempt at a definition of a flood. For example: ". . .harmful inundation of property and land utilized by man"; it is difficult to imagine what harm*less* inundation is. Even better, "a flood is a body of water which rises to overflow land which is not normally submerged". Well I never.

Worse than the floods of October 1993 were those of October 1960, when a very wet summer was followed by a very wet September and some record-breaking rains in October. Indeed these were arguably the most disastrous autumn floods within living memory – more extensive and longer lasting than the Home Counties flood of September 1968. Devon was probably the worst-affected county and Exeter had five separate floods in nine weeks.

According to the relevant volume of *British Rainfall* the first West Country flooding hit Cornwall on 28 September, with over twenty people rescued by boat from upstairs windows in the villages of St Columb Major and St Mawgan. On 29 and 30 September over five inches of rain fell in south Devon, and Exeter was temporarily cut off from the rest of the county with every major road out of the city under water. On 6 October a thousand homes were flooded in Exmouth, and much of the equipment used to clear up the mess following the previous storm was lost.

A week of cold sunny weather mid-month provided something of a respite, but torrential rain returned on 26/27 October, and floodwaters again swept through the middle of both Exeter and Taunton. The last serious flood of the season caused widespread damage in Glamorgan on 3 December, following six inches of rain in the Brecon Beacons above the heavily populated Welsh valleys. There were disastrous floods in the valley of the River Ogmore, and water in the town centre at Bridgend was reported to be three feet deep.

Away from the southwestern corner of the country, a summer-type thunderstorm struck the small town of Horncastle in Lincolnshire on 7 October. In a little over five hours, 7.24 inches of rain fell – that is roughly four months' worth of rain during a single afternoon. It is believed that most of the rain fell within three hours at the height of the storm. At the nearby Revesby reservoir, the water level rose three feet in as many hours, machine houses were flooded out, and local rivers broke their banks. In Horncastle itself, the main streets were turned into torrents of water carrying household furniture and large quantities of stock from the shops in the town centre, and cars and caravans floated away too. The lower part of the town was under six feet of water, and one inhabitant was drowned.

Taking a climatological view, October 1960 had been a very interesting month. The repeated downpours in the West Country were caused by active Atlantic depressions heading into the Southwest Approaches and then coming to a grinding halt, and thus depositing their contents at their leisure. Now, persistent low pressure near southwestern Britain means that the rest of the British Isles will lie under the easterly airflow on the northern flank of the depressions, with relatively high pressure located to the north of Scotland. That sort of arrangement is unusual at any time of the year, but particularly in October. On this occasion, prolonged easterly winds over northern Britain meant that the Highlands and Islands of Scotland enjoyed an exceptionally dry and reasonably warm October, in complete contrast to areas further south. Thus while Princetown on Dartmoor recorded 15 inches of rain during the month, parts of the Isle of Skye received less than one inch. Similarly, Horncastle had more than five times its normal October rainfall, while Fort William had little more than 10 per cent of its local average, making it one of the driest Octobers of the 20th century there.

13. November

"No sun, no moon, no morn, no noon,
No dawn, no dusk, no proper time of day;
No sky, no earthly view, no distance looking blue,
No road, no street, no 't'other side the way';
No end to any row,
No indication where the crescents go;
No top to any steeple,
No recognition of familiar people.
No courtesies for showing 'em,
No knowing 'em,
No travelling at all, no locomotion;
No inkling of the way – no motion;
'No go', by land or ocean –
No mail, no post,
No news from any foreign coast;
No park, no ring, no afternoon gentility;
No company, no nobility;
No warmth, no cheerfulness, no healthful ease;
No comfortable feel in any member;
No shade, no shine, no butterflies, no bees,
No fruits, no flowers, no leaves, no birds . . .
November!"

THOMAS HOOD (1799–1845)

Thomas Hood's vivid evocation of typically drab foggy November weather – and the gloomy cast of mind that follows from it – reminds us that these days we distance ourselves from our natural surroundings and that we rarely give ourselves the opportunity to absorb the character of the seasons, especially during the colder half of the year. We now live in well-lit centrally heated homes, and travel to our well-lit centrally heated work-places in warm comfortable cars or reasonably efficient (well, sometimes) public transport. When Hood wrote his words our forebears perhaps had oil lamps, candles, coal fires and warming pans, and that's all. They could not have imagined the extent to which we, a century and a half later, can ignore the unfriendliness of late-autumn and winter weather.

There are perhaps two standard images of November. One is of endlessly grey gloomy days, a dank fog hanging in the middle distance, and the autumn leaves lying damp and inert. The other is quite different. It is of wild stormy days, rain beating on the windows, trees bending in the wind, brown leaves swirling, and people struggling against the odds to keep umbrellas under control. Both of these images form important aspects of November weather, but rarely does one of these patterns persist throughout the month.

Sometimes November surprises us, and produces a spell of mellow autumn sunshine, or a sudden burst of northerly winds may bring several days of crisp bright weather with cloudless skies and almost unbroken winter sun. A study of November sunshine statistics reveals some interesting trends. Indeed one of the most notable changes in the British climate in the 20th century has been the improvement in sunshine records during this month. A large part of this improvement has been the result of the reduction in the amount of smoke that used to pollute the air of our towns and cities. Thus the increase in sunshine hours has been greatest in the middle of our conurbations. But rural areas have become a bit sunnier too, and this is the result of a change in weather patterns in November since about 1965, with a marked increase in the frequency of relatively clear northerly and north-westerly winds, and a decrease in cloudy southerlies and easterlies.

Daily records of sunshine were kept at Kew Observatory, in southwest London, from the mid-1870s until 1980 when the Observatory closed, and subsequently next door at Kew Botanic Gardens. Between 1884 and 1964, not one November managed so much as 80 hours of bright sunshine, but in the next thirty years there were no fewer than nine Novembers with 80 hours or more, and in 1989 the month's total was a phenomenal 104 hours – roughly the normal for October.

The 30-year standard periods for Kew gave the following results:

1881–1910	51 hours
1891–1920	52 hours
1901–1930	53 hours
1911–1940	52 hours
1921–1950	52 hours
1931–1960	53 hours
1941–1970	58 hours
1951–1980	66 hours
1961–1990	69 hours

and the figure for 1964–1993 had reached 70 hours – an increase of 33 per cent in little more than 30 years.

Because of its location in the London suburbs, Kew's figures will have benefited from the improvement in air quality during the 1950s and 1960s, but much less so than central London sites. At Kingsway for instance the 1921–1950 average was 38 hours compared with 70 hours for 1961–1990, an increase of 84 per cent. The changes in rural areas, clear of London's smoke, were smaller but still significant. For example, Rothamsted (25 miles north of London in the Hertfordshire countryside) averaged 61 hours and 66 hours for the respective periods, an increase of 8 per cent.

There is no reason to suppose that the increased dominance

of northerly and northwesterly weather types which began in the 1960s will be maintained indefinitely. There have been changes in the character of other months that have lasted several decades and then ended for no immediately apparent reason: for instance the marked warming of October between the 1920s and the 1960s ended abruptly around 1973–74. Thus dismal, dreary Novembers may yet become the norm again. But at least our city centres will remain much brighter at this time of the year than they have been for centuries.

PURPLE PATCHES IN A DRAB MONTH

Even the sunnier Novembers could hardly be called colourful, and dingy greys and browns are probably the hues most of us would associate with this month. It is not really the time of year you would normally associate with bright colours and dramatic sky phenomena, although one or two eye-catching sunsets – crimson or coppery, say – would not be out of place on fine frosty afternoons.

Most of us will stop to admire a vivid sunset or sunrise, but other colours that occur in the sky from time to time are hardly ever noticed. Under certain atmospheric conditions with unusually high levels of moisture, the blue sky may appear rather more green than blue, while a brilliant purple glow is a well-documented but comparatively rare visitor to this country.

The most striking occurrence of purple sky colouring in the United Kingdom took place on 27 November 1979. The *Daily Telegraph* the following day reported that "the evening sky turned bright pink, then mauve, deepening into purple, over a large area of the Southeast and East Anglia. . ." For some five minutes the whole sky was suffused with this gawdy purple glow, and I well remember people coming out of their houses, staring and pointing and generally bewildered, on that abnormally mild afternoon. Subsequently the eerie colours vanished very rapidly as the light dimmed. Later reports showed that this unusual phenomenon had not been confined to England, having been noted in newspapers in the Low Countries and in Germany.

The causes of this particular event lay partly in the configuration of cloud layers just after sunset, and partly in the existence of a cloud of dust high in the atmosphere. This dust cloud had originated over the western Sahara some three days earlier, when vigorous winds and powerful convective currents had lifted enormous quantities of sand and dust to an altitude of up to five miles. The larger sand particles had fallen out quite quickly under the influence of gravity, but prevailing winds aloft brought the cloud of fine dust northwards across Portugal and the Bay of Biscay to the British Isles, and then eastwards to other parts of northern Europe. Later, sporadic outbreaks of rain brought dust falls to many parts of Britain and adjacent parts of the continent. According to press reports and other published accounts, the heaviest dust falls in the United Kingdom occurred in the Lancashire and Merseyside areas, where yellowish-white material was brought down by the rain in the Liverpool and Blackpool areas, leaving muddy splashes on motor vehicles and ruining lines of washing. Heavy falls were also observed in the Irish Republic, notably on the cities of Cork and Dublin, and some of the accounts from Ireland spoke of red dust as well as the lighter colours. One local scientist calculated that some 60,000 tons of dust fell in the Cork region alone.

These dustfalls occurred variously between 28 and 30 November, but chiefly on the 29th. However, during the late afternoon of 27 November the main effect of the dust cloud was to produce a particularly vivid red and crimson sunset.

The two main layers of cloud were crucial to the appearance of purple rather than red light. The upper layer of altocumulus was at an altitude of some two to three miles, and was illuminated by the setting sun across practically the whole breadth of the sky. At a much lower level, just a few hundred feet above the ground, there was a tenuous sheet of stratus cloud, already in the Earth's shadow, and it therefore appeared to be dark blue-grey in colour. Thus the bright pinks and reds filtering through this much darker but very thin cloud sheet were transformed into mauves and purples. The colour persisted until the upper cloud sheet itself passed into the earth's shadow.

This dramatic sort of purple glow should not be confused with "purple light", which is a very much subtler colouring of a cloudless evening sky that occurs regularly after sunset under very clear atmospheric conditions.

An earlier example of a similarly exotic sunset occurred way back in September 1950, when the culprit was traced to widespread forest fires which had been raging for several days before the event. That occasion famously produced the last significant example of a blue moon (and indeed a greenish-blue sun) to be observed in Britain. There were no such reports after the November 1979 event – or at least none published in the national newspapers or the meteorological literature.

STORIES OF DUCKS, SLUSH, MUD, AND LONG-RANGE FORECASTING

It has long been held by country folk that November weather is a sure pointer to the sort of winter to come. Richard Inwards' classic volume *Weather Lore*, first published in 1869, contains several ancient sayings that bear this out. Probably the best known of these old saws is one that runs:

> If November ice will bear a duck,
> There'll be nothing after but sludge and muck.

This seems to imply that a cold November (or at least one with a spell of severe weather) will be followed by a mild and rainy winter. The complementary proposition is supported by the following saying, which regrettably does not contain such a memorable rhyme:

Flowers in bloom late in the autumn
Indicate a bad winter to come.

You may still hear these saying trotted out on appropriate occasions by the ants, moles and fir cones brigade of amateur weather prophets, but not even the most liberal professional has any time for them, simply because the facts do not bear them out. In the last 50 years or so, there have been 7 notably cold Novembers, 2 of which were followed by mild winters, 3 by average winters, and 2 by cold winters. Looking at it the other way round, the 16 very mild winters that have occurred since 1925 were preceded by 1 mild, 12 average, and 3 cold Novembers.

One of the coldest Novembers in the last half century happened as recently as 1993. Watching the television weather forecasts during that month, you could have been forgiven for thinking that we had never had a cold snap in November before, let alone a few snow flurries. Certainly much of the UK – and indeed a large portion of continental Europe – endured a noteworthy early spell of wintry weather, but it would be wrong to say that severe frosts and snow are rare during the second half of November. Indeed, the temperature has been known to fall as low as −23 °C (−10 °F) in Scotland as early as 14 November, as happened in 1919. In 1993, the lowest temperature in Britain was −14.8 °C (5.4 °F) at both Braemar and Grantown-on-Spey on the 24th, while on 30 November 1985, the temperature fell to −20.9 °C (−5.6 °F) at Kinbrace in Sutherland.

The average number of November mornings with snow on the ground is greatest in upland areas of Scotland, with seven at Braemar and five at Aviemore. There is one such morning in Edinburgh and Glasgow, York and Lincoln, but a November snow cover occurs only once every three years in Oxford, Cambridge, and Birmingham, once every five years in the London suburbs, once every ten years in Manchester, and more infrequently even than that in central London and in Cardiff and Plymouth. Thus it can be seen that many years may pass by without any serious wintry weather in southern and southwestern England and south Wales in November.

Serious snow last occurred over a large area during this month in 1993, and since the last war widespread disruptive snowfalls have happened during the Novembers of 1988, 1985, 1980, 1973, 1972, 1969, 1965, 1962, 1952, and 1947.

In 1993, snow fell over most eastern and Midlands counties of England and also across eastern and central Scotland between 20 and 22 November, reaching a depth of eight inches in the Aberdeen and Royal Deeside areas, and two to three inches in East Anglia and Lincolnshire.

In 1988, 20 November brought substantial snow across a broad belt extending from central Scotland to Kent, generally to a depth of two inches or so, although a foot of snow lay for a time over the hilly interior of the Isle of Man. Further snow fell over the next two days in Kent, bringing the depth over the Downs south of Canterbury to six inches, and Dover was cut off from the rest of the county for a time.

During November 1985, snow lay for 14 days on Speyside and Royal Deeside, for 9 days in Aberdeen, 6 at Huddersfield, and 4 at Birmingham and Norwich. On the 18th snow even fell for several hours in central London. The end of the month was exceptionally wintry in Scotland, with the very low night-time temperatures noted above, and eight inches of snow blanketed Glenlivet in Banffshire.

The heaviest November snowfalls in living memory occurred in 1965 and 1952. In November 1965 the ground was snow-covered on 18 days at Braemar, and for 10 days as far south as RAF Waddington, near Lincoln. By the end of the month snow lay almost two feet deep in the environs of Durham. Late in November 1952, heavy snow fell on four consecutive days in a broad band across south Wales, the Midlands and East Anglia, with many places reporting a blanket of snow at least six inches deep. The eminent amateur meteorologist E.L. Hawke ran a weather station at that time at Whipsnade, high in the Bedfordshire Chilterns, and he measured level snow ten inches deep on the morning of the 30th, with drifts eight feet high in the village.

Just out of interest, the winters of 1952–53 and 1965–66 were both average to rather cold, and both contained spells of very cold wintry weather in each of three main winter months.

THE COLDEST NOVEMBER OF ALL

At 11 o'clock in the morning on 11 November 1919, the whole of Britain observed a two-minute silence, marking the first anniversary of the armistice. This day also ushered in an extraordinary spell of wintry weather, the like of which has not been seen in November since. Indeed, the cold was so intense and the snowfalls so heavy that this particularly cold snap ranks alongside the severest weather in any 20th-century January and February – including the historic severe months of February 1947 and January 1963.

Even before November arrived, 1919 had been a most unusual year, weatherwise. A cold winter and early spring were followed by a prolonged heatwave in May and early June, but northerly winds set in during the second half of June and lasted for some six weeks, resulting in the coldest July of the century. Hot sunshine returned during August and early September, but a further dramatic change in fortunes brought unprecedentedly early snowfalls around 20 September – just nine days after the hottest day of the year. Autumn 1919 was characterized by repeated surges of arctic air in the form of strong northerly winds, and following the September snow flurries, October was one of the coldest of all time, with frequent severe frosts. Thus was the scene set for the wintry November.

According to the account of the month's weather in *Symons's Meteorological Magazine*:

> The month opened with cold easterly winds, but it was not until after the 7th or 8th that the thermometer descended to an excep-

> tionally low level. The sharpest frosts appear to have occurred between the 12th and 15th, when the sheltered thermometer fell below. . .10 °F in Scotland.

When all the returns had been received by the meteorological authorities, it was clear that in a few localities temperatures had fallen very much lower than this, especially overnight on the 13th/14th. Lowest of all was a reading of −23.3 °C (−10 °F), logged at Braemar, the notorious cold spot in the upper Dee valley in the Grampians. This figure remains to this day the lowest ever recorded in November in the UK by a wide margin. This is of course all the more remarkable for its occurrence during the first half of the month. Even in December, January and February, lower temperatures have occurred since then only in the winters of 1954–55, 1978–79, 1981–82, and 1983–84.

If proof were needed that the Braemar figure was authentic, −21.7 °C (−7 °F) was recorded at Perth, and −21.1 °C (−6 °F) at both Balmoral, just down the road from Braemar, and at West Linton, which is some 15 miles southwest of Edinburgh. At Balmoral, the maximum temperature on the 14th was a phenomenal −10 °C (14 °F). The arctic weather's grip over England and Wales was a little less powerful than it was over Scotland, but some notable readings were obtained south of the border as well. Lowest in England was −12.7 °C (9 °F) at Scaleby, near Carlisle; in Wales the temperature fell to −9.4 °C (15 °F) at Rhayader, in Radnorshire; in Ulster the lowest reading was −12.2 °C (10 °F) at Lisburn; and in the rest of Ireland −11.1 °C (12 °F) was logged at Markree Castle, County Sligo.

Snow fell intermittently over much of the country from 8 November onwards, but to no great depth until the 11th. However, on the night of the 11th/12th and throughout the following day a severe snowstorm raged, depositing 8 inches in the streets of Edinburgh, 12 inches over Dartmoor, and 17 inches at Balmoral, where snow remained on the ground throughout the remainder of the month.

And just for once, an icy November was followed by a sludgy, mucky winter. Mean monthly temperatures during December 1919, and January and February 1920, were well above average in all three months. December was also a very wet one, although by February the weather had become mostly dry and settled.

MOTORWAY MADNESS

It is November, and naturally enough fog appears on several mornings during a typical month. Rarely does a foggy spell pass by without a number of serious road accidents, and every so often a major pile-up will happen on one of our motorways or other major trunk roads. "Motorway Madness", the police authorities dubbed it when a series of disastrous accidents occurred in the late 1960s and early 1970s. And the name has stuck.

I once heard a radio presenter introducing an item about motorway madness say that the crash had been "caused by thick fog". This is a trifle unfair to the fog. Crashes in these conditions are caused by human beings driving their vehicles too fast and too close together, and simply unable or unwilling to accept that they have to adjust their driving pattern when the visibility is very poor.

Meteorologists define fog as occurring when horizontal visibility falls below 1000 metres (1100 yards), but most ordinary people would describe the weather as simply "very misty" with visibility between, say, 500 and 1000 metres. Motor vehicles should use headlights under such conditions, but there is little need to reduce speed substantially until visibility deteriorates to roughly 250 metres. Indeed, for official meteorological purposes there is a 200-metre (220-yards) threshold that separates "fog" from "thick fog", and another at 50 metres (55 yards) at which point "dense fog" is deemed to begin.

It is worth pointing out here that nearly everybody seriously exaggerates the density of fog when they are asked to estimate the visibility. This includes traffic reports on radio and television, which often depend on information passed on by the police, or by motoring organizations such as the AA and the RAC. Occasionally you will hear mentions of "nil visibility", which is certainly something I have never experienced – in fact on such occasions a trained meteorological observer will probably report something between 25 and 50 metres. For myself I have never experienced horizontal visibility at ground level below about 20 metres.

I tried a small, not terribly scientific, experiment one foggy morning in December 1991. Having assessed the horizontal visibility at approximately 150 metres (plus or minus 20 metres), I asked a dozen journalists, broadcasters and administration staff from the independent radio station LBC to make a serious estimate of the distance at which they could distinguish the outline of buildings and vehicles. Of the 12 estimates, all were between 50 and 125 metres. In other words, everyone underestimated the visibility – by between 16 and 66 per cent.

So, fog can be categorized by its density, but the character of fog can vary enormously in other ways as well. As far as driving is concerned, the most dangerous types of fog are freezing fog and drifting fog patches.

Freezing fog is something of a misnomer. At temperatures below freezing point, fog droplets are still liquid, and are described as being "supercooled". However, on impact with a solid object these minute water droplets will freeze immediately, leading to a build-up of ice. Thus it is essential to keep the car windscreen supplied with warm air so that the external temperature of the screen remains above freezing. On busy motorways such weather often produces a lethal spray containing water, mud, oil, and perhaps also salt, and this concoction rapidly reduces windscreen visibility, and it is only then that you realize that the washer nozzles have frozen up.

The majority of serious motorway pile-ups occur when there are dense patches of fog drifting across the road. With blue sky and bright sunshine the motorist may have a false sense of security, only to run into a "wall" of fog lying in the next dip the road. A low sun shining obliquely onto the fog bank increases the apparent opacity of the fog.

One telling statistic: some 5 per cent of motorway casualties occur during thick fog, but in an average year thick fog occurs for less than 1 per cent of the time.

THE WETTEST MONTH OF THE YEAR

Although some Novembers are quiet and foggy, such months are definitely in a minority. November is probably more likely to be wet and stormy than any other month of the year, while the frequency of fog in most parts of Britain is actually a little higher in December and January.

The first systematic rainfall measurements in Britain were made over 300 years ago, and since 1727 there has been a sufficient scatter of rainfall records across England and Wales to enable a "national" monthly rainfall total to be calculated. When averaged over a long period, these watery statistics confirm that November is consistently the rainiest month of the year – at least south of the Scottish border. The England and Wales average rainfall for the month is 98 mm (3.9 inches), some 7 per cent higher than December and January, which are the next wettest months, and 70 per cent higher than April, which is the driest.

Even when the statistics are broken down into the five main regions (Northeast, Northwest and North Wales, Midlands and East Anglia, Southeast, Southwest and South Wales), November is still the wettest month. Only when we look on a more localized basis do we find exceptions. For example, in parts of Cornwall December is marginally wetter, on Tyneside August is the wettest month, while locally in East Anglia November comes third after July and August. In Scotland, broadly speaking, July is the wettest month of the year in the eastern half whereas December is wettest in the western half.

It is, of course, impossible to provide a simple explanation why November is so wet, but we can identify some of the influences. At this time of the year the land masses of Asia and North America are rapidly cooling down, while the oceans cool off much more slowly, and the increasing temperature contrast between continent and ocean fuels the big Atlantic depressions that bring so much of our rain. Meanwhile, temperatures are higher in November compared with December and January, especially over the seas surrounding the British Isles; and the warmer the air is, the more moisture it can hold, and therefore the greater the potential for rain.

The unsettled nature of November's weather is further illustrated by the fact that four of the five wettest months of the last 230 years were Novembers (the other was October 1903), while there has been a notable absence of very dry Novembers. The November of 1945 was the driest over England and Wales with a national total of 17 mm (0.7 inches), ranking only 71st driest of all months.

It is not just Britain where November is the wettest month of the year – the same is true over large swathes of Europe, in particular the Mediterranean lands, where the comparatively warm waters of the Mediterranean Sea provide copious moisture. In Rome, for instance, the November average is 129 mm (5.08 inches), more than an inch higher than the next wettest month.

Italy's worst floods of the 20th century occurred in November 1966. The country was virtually cut in two and most transport arteries between the northern plain on the one hand and Rome and the Mezzogiorno on the other were impassable for several days. The death toll rose to 113, most of victims being drowned. The human catastrophe was amplified by what was arguably Europe's greatest cultural disaster since the war as floodwaters rose to unprecedented levels in Florence and Venice, inflicting untold damage. According to one chronicler of events, in Florence

... millions of rare books, 750,000 valuable ancient letters and manuscripts, and 1300 paintings and etchings were damaged or destroyed. In the Biblioteca Nazionale 1.3 million volumes were damaged; at the Library of the Jewish Synagogue some 14,000 books were ruined; at the Gabinello Vieusseux a quarter of a million were lost, as were 36,000 at the Geography Academy. At the Music Conservatory the entire collection was demolished. The thirteenth-century church of Santa Croce had to be completely rebuilt, as was the San Firenzi Palace. The frescos in the Medici Chapel began to blister and peel, while six hundred paintings were under water for several hours in the famous Uffizi Gallery.

14. December

"A green Christmas makes a fat churchyard."
ANON

There was a time, probably right up to the early decades of the 20th century, when mild winters were feared. They were regarded as a breeding ground for germs, leading to rising mortality rates, thence inevitably to those "fat" churchyards. A more irreverent version runs something like this:

For a warm Christmas Day,
The parson doth pray.

This particular saying acknowledges the fact that the local clergyman supplemented his living with burial fees – the warmer the winter, the higher the death rate, and the fatter the parson's pocket.

The tradition of white Christmases probably owes more to Charles Dickens and Dutch painters of the 17th and 18th Centuries than it does to anyone or anything else. At no period in the last few centuries has the Christmas period been a particularly snowy one, and Christmas card scenes are very much the exception rather than the rule.

These days, the bookmakers have hijacked the white Christmas. It is one of the more bizarre aspects of reporting the weather in the 1990s that we can have a foot of snow on the ground on Christmas Day but we may not call it a white Christmas. The official statistics have been geared to the requirements of Messrs Ladbroke, William Hill, et al., and they only pay out if flakes of snow are observed falling on what used to be called the Air Ministry Roof (or regional equivalent). It need only be a fleeting flurry at 40 Fahrenheit, or a couple of blobs of sleet on an otherwise grey rainy day. But snow on the ground simply does not count.

The most recent Christmas to make the grade was in 1993. And because it met the requirements of the bookies, we were told it was the first white Christmas since 1976 by the news on radio and TV. It is true that snowflakes fell over roughly 40 per cent of the UK that day, but in many places it was literally just a few flakes. Most of Wales together with parts of the west Midlands and the West Country had some rather more persistent snow, but very little settled on the ground except over the hills. Amongst our major cities, only Aberdeen, and the hillier suburbs of Edinburgh, Greater Manchester, Liverpool, Bradford, Leeds and Sheffield had any significant snow on the ground. Thus, between 3 and 4 million people in Britain woke up to a white Christmas morning, which means that 54 million didn't!

There were plenty of pictures of "snowbound Britain" in the television news bulletins. But did you know that on occasions like this news editors ring up the weather forecasters to ask where they should send the camera crew for a good Christmas-card snow scene?

Christmas 1993 joins a short list of white Christmases which, by any common-sense definition, weren't. Christmas 1976, for instance, was a fairly sunny day with a chilly wind, but it was not really all that cold. However, a five-minute-long flurry of sleet was reported by the official weather observer in central London at about one o'clock in the morning, so it counted. Christmas 1964 brought scattered snow showers to some east-coast counties of England together with parts of Scotland and Northern Ireland, but the greater part of the UK was fine and sunny.

At the other extreme, 1981 was the last occasion most people had a truly white Christmas with snow lying – fairly deep, quite crisp, and relatively even. This was thanks to one of the coldest and snowiest Decembers of the 20th century. Heavy snowfalls had occurred frequently from 8 December onwards, and indeed many places reported snow falling daily from the 19th to the 24th inclusive. But no snow fell on the 25th, so the bookies hung on to their dosh.

Prior to 1981, the last generally snowy Christmases were in 1970 and 1968. In 1970 snow fell intermittently throughout the day over much of England, but it was mostly light, and few areas ended the day with more than an inch of two on the ground. Most of the West Country and much of Scotland escaped completely. In 1968 heavy overnight snow in the Midlands and Wales had stopped by first light on Christmas morning, leaving a blanket of snow over a foot deep in the Welsh Marches and almost as much in the Cotswolds. Before that, Christmas Day 1962 was snowy in parts of Scotland, and the rest of Britain joined in on Boxing Day; a large chunk of England remained snow-covered until the first week of March 1963.

Probably the best example of Christmas snow this century was in 1938, when 6 to 12 inches fell widely on 21–22 December, and stayed for a week. The snow was dry and powdery, settled on roads even in central London, and drifted gently in the moderate winds. But the most widely quoted instance, that of 1927, didn't really get going until Christmas night. However, it turned out to be one of the worst snowstorms of the last hundred years, and most of the country awoke on Boxing Day to find themselves snowed in – drifts 10 to 20 feet deep were widely reported.

But between these rare snowy examples, the vast majority of Christmas Days have been relentlessly green. And in some parts of the country snow is more likely to fall on Easter Day than it is on Christmas Day.

THE GALE OF DECEMBER 8/9, 1993

It was certainly a wild and windswept night. The telegraph wires whistled, the wind rumbled unnervingly in the chimney, and countless dustbin lids rattled noisily around countless backyards. The weather forecaster sharpened his 4B pencil so he could cram in all those extra isobars, cowboy builders gleefully rubbed their hands in the expectation of a bonanza of roof repairs, while insurance-company executives also had a collective gleam in the eye – here was a good excuse for another hike in premiums for the following year. And the next morning weather historians pored over dusty tomes in their forlorn quest to satisfy the insatiable appetite of radio, TV and the newspapers for records and statistics. It was something the British media do so terribly well – a Great British Weather Disaster.

They happen about once a month. In November 1993, you might remember, we had all that unseasonable ice and snow and freezing fog (heavens above, it was only November), and in October a fortnight of rain led to widespread flooding in eastern England. Admittedly, Wednesday night's gale was arguably the most dramatic weather event of 1993, on a par with the disastrous June floods in Wales and the West Country in terms of financial cost, and exceeding anything since the Burns' Day Storm of 1990 in terms of human lives lost – on that fateful day the death toll reached a dreadful 48.

Severe gales have been part and parcel of the climate of northwestern Europe since the beginning of history. The eminent climate historian Professor Hubert Lamb has spent the best part of two decades researching them, and the result of his great scholarship is a volume entitled – prosaically enough – *Historic Storms of the North Sea, British Isles and Northwest Europe*. In it he chronicles in great detail probably every major gale and North Sea flood during the last 300 years, with a more sketchy outline of the most notable disasters in the region during several preceding centuries. The storm and flood of January 1362, known as the

"Great Drowning", is considered to be have been the region's greatest natural catastrophe of the last thousand years. Vast tracts of what are now Denmark and northwest Germany were inundated, and roughly half the local population were drowned. Contemporary accounts put the death toll at 100,000, although Lamb suggests a figure below 30,000 is more likely. Whatever the figure, this was a formidable disaster coming just a few years after the Black Death. In England the storm caused widespread damage, and church towers were demolished in London, Norwich and Bury St Edmunds. The Great Storm of 26–27 November 1703 (on the old Julian calendar) took 8000 lives, including a substantial proportion of Britain's naval and merchant fleets, while the infamous North Sea flood of 31 January 1953 drowned some 2000 in the Netherlands and over 300 in eastern England.

These days, whenever we have a newsworthy weather event in this country, we seem to think that it is something brand new. Records and statistics are paraded, often out of context, to "prove" that it has been hotter or colder or wetter or windier than ever before. The December 1993 gale, although destructive, was not exceptional by any reasonable measure. Highest gusts recorded that windy night were just shy of 100 mph – several reports of 97 or 98 mph coming from south Wales, the West Country, and the Kent coast – but similar wind speeds had been recorded over southwestern Britain during the preceding January, while in Scotland gales were even more severe. The Burns' Day Storm three years before that brought gusts of over 100 mph in Sussex and Kent and 108 mph at Aberporth in Dyfed and Gwennap Head in Cornwall.

Some people also seek to apportion blame: "It must be the greenhouse effect . . . or the ozone hole," they say. These are the latest in an honourable list of culprits that have been trotted out over the years to explain away bad weather. In 1980 the eruption of Mount St Helens in the state of Washington in the USA was blamed for the appalling summer that followed in Britain and

Europe (see Chapter Nine). In the 1970s it was supersonic aircraft, in the 1950s nuclear testing, in the 1920s radio, and in 1917 – a notoriously wet summer – it was gunfire in Flanders.

I wonder who the poor soggy survivors of the 1362 flood blamed?

THE GREAT LONDON SMOG

Conscious as we are of our environment these days, and especially of the pollutants we continue to pump into the atmosphere, no one who has lived in London or Manchester of Glasgow for forty years or more could deny that there has been a substantial improvement in the air quality in our big cities since the 1950s. Nevertheless, while soot and sulphur dioxide have decreased, it has to be pointed out that other less visible but still toxic chemicals have increased, and continue to do so.

It was the Clean Air Act of 1956 which, after 650-odd years of piecemeal and largely ineffectual legislation, succeeded in cleaning most of the smoke out of our city air. The main target was the ordinary coal fire in the ordinary home, and an 80 to 90 per cent reduction in smoke emissions was achieved by the introduction of smokeless zones and smokeless fuels.

It is not surprising that the Clean Air Act was given teeth, because the London smog of December 1952 – the smog that precipitated parliamentary action – was the single most deadly meteorological catastrophe in Britain for approximately 250 years. Most reference books tell us that 4000 people died as a result of the 1952 smog, but this is inaccurate. Reference to the statistics published by the health authorities shows that mortality rose by 4000 during the first half of December 1952 in the old London County Council area alone, and in fact a further 2000 deaths in the suburban areas of Middlesex, Essex, Kent, Surrey and Hertfordshire were attributable to the smog. Most of the people who died were among the elderly and chronically ill; the death rate from bronchitis was ten times the December average, influenza seven times, and pneumonia five times. Hypothermia no doubt contributed to the total as temperatures remained below freezing thoughout, though this was rarely given as a cause of death in 1952.

For some 60 hours at the peak of the smog, visibility was officially reported at 22 yards or below at various monitoring points within Greater London, and there were several reports of 10 yards or below. London's traffic was severely disrupted throughout the five days, air and river transport were halted, and football matches were postponed. Smoke and sulphur dioxide concentrations were greatest in Lambeth, Westminster, Southwark and the City, where daily values averaged 10 to 20 times normal background levels throughout the foggiest period. But over a short period at one individual monitoring point it was estimated that smoke density was between 50 and 60 times the December normal – that is, the normal for the early 1950s.

In the aftermath, the Conservative government of the time did not display any keenness to consider legislation, in spite of some very noisy ministerial question times when Parliament returned after their Christmas holiday. Initially they preferred to highlight the powers that local authorities already had, but ministers eventually succumbed to pressure and set up a committee in 1953 (the Beaver Committee) to report on urban pollution. Parliamentary committees and Royal Commissions usually represent attempts at procrastination by governments, but if this was the case here it was foiled by the Conservative MP Gerald Nabarro, who introduced a private member's bill. The government finally introduced its own Clean Air bill in 1955, though it was delayed by the general election of that year, but the Clean Air Act finally reached the statute books on 5 July 1956.

The Act was not as powerful as many would have liked, with legal controls only on smoke emissions, and it also allowed a seven-year transition period before full compliance was required. But considerable improvements were plain for all to see by the

1960s. This was just as well, because in early December 1962 – 10 years to the very week after the Great Smog – another smog developed over the capital. On this occasion the fog was just as thick and lasted even longer, and peak sulphur dioxide levels were actually 10 per cent up on the 1952 event. But smoke concentrations had dropped by an astonishing 60 per cent, and the excess mortality (that is, the increase in number of deaths over and above the normal number) was just 700 in the London County Council area – a drop of over 80 per cent.

London suffered more than other urban areas because of its size and population. Having said that, serious smogs occurred in Glasgow and Clydeside in November 1925 and November 1977, in Greater Manchester in January 1971, and in both Manchester and Birmingham in November 1936.

In recent years, although the classic smogs of earlier decades no longer occur, there has been a vast increase of less visible pollutants from vehicle exhausts and power stations – particularly carbon monoxide and nitrogen oxides – which remain a serious danger to health. This is particularly so under the kind of weather patterns that prevailed in December 1952. Notable recent examples occurred in November 1988, later November and early December 1989, and mid-December 1991 – all of which were accompanied by an increase in chest problems, particularly among the very young and the elderly.

WINTER WARMTH BROUGHT BY BALMY BREEZES FROM BERMUDA

The shortest day of the year – the winter solstice – usually falls on 21 December although it can also occur on the 20th or the 22nd thanks to the leap-year cycle. Not that you will notice much difference in day length between early December and mid-January, while for a week before and after the solstice the length of daylight varies by less than three minutes. The sun is above the horizon for 7 hours 49 minutes in London on the shortest day, but in the Channel Islands it is approximately 8¼ hours, while in Shetland it is barely six hours. And of course the sun is at its lowest in the sky at this time of the year – a mere 6 degrees above the horizon at midday at Lerwick, but 15 degrees above in London, and almost 18 degrees above on Jersey.

These figures emphasize what is common sense: that the midwinter sun is very weak and is likely to have comparatively little influence on temperature, even on the brightest of days. This means that December's highest temperatures in the United Kingdom are almost exclusively the result of imported warmth. The main factors, therefore, are the prevailing temperature where the air mass originates, the modifying influences on the journey between the air mass's origin and Britain, and the speed with which the air travels from its source region. All warm air masses will be cooled to some extent on their journey, even if it is a rapid one across the relatively warm waters of the Atlantic Ocean. Thus our highest temperatures at this time of the year are likely to be accompanied by strong winds, thick cloud, and a moist atmosphere.

Most Decembers will contain at least one day with a strong-to-gale-force southwesterly wind bringing air from subtropical latitudes over the Atlantic Ocean, probably somewhere near the Azores, and temperatures will climb to the mid-50s Fahrenheit (around 13 °C). More rarely, when the configuration of lows and highs is just right, an airflow will reach the British Isles directly from the region southwest of the Canaries, or perhaps from Bermuda, and on these occasions the temperature has been known to exceed the 60 Fahrenheit mark (over 15.5 °C). Such a surge of warm tropical air flooded across the country on 18/19 December 1993, lifting the temperature to exactly 15.5 °C (59.9 °F). Two years earlier, on 22 December 1991, the mercury climbed to 15.6 °C (60.1 °F) at Elmstone, near Canterbury.

But all this does not explain the very highest readings observed in Britain in December. A clue lies in the locations at which these abnormally high temperatures occur: 18.3 °C (65 °F) at Achnashellach in Wester Ross on 2 December 1948; 18.0 °C (64.4 °F) at Aber, near Bangor, in Caernarvonshire on 18 December 1972; and 17.7 °C (63.9 °F) at Cape Wrath, the northwesternmost tip of Scotland, on the same date. More recently, a whole swarm of exceptional temperatures occurred on 1, 2 and 3 December in 1985, including 17.2 °C (63.0 °F) at Bude in Cornwall, and 16.7 °C (62.1 °F) at Cardiff Airport. All these locations are to the lee of high ground when the wind is blowing from the south. They benefit from a process known as "adiabatic compression", which occurs when air from well above the level of the mountains descends to much lower levels, and the higher pressure near sea level compresses the air resulting in a rise in temperature.

Adiabatic compression is more popularly known as the "föhn effect" – the föhn wind is an Alpine phenomenon that brings abnormally high winter temperatures to Alpine valleys in Switzerland and Austria when the large-scale windflow is from the south. On rare occasions, a northerly föhn brings a dramatic rise of temperature to the north Italian cities of Turin and Milan, courtesy of very mild Atlantic air sweeping eastwards from the Bay of Biscay across northern France, then curving southeastwards into southern Germany, and southwards across the Alps. The Lombardy Plain in winter is often filled with stagnant, cold, foggy, polluted air with temperatures often staying near or below freezing point, but the arrival of a northerly föhn can bring bright sunshine, a stiff north to northwest wind, and temperatures of 15–16 °C (around 60 °F).

Equivalent effects are found elsewhere in Europe, too, wherever warm moist winds cross mountainous regions. The "vent d'autan" is such a wind in the southwest and west of France where the Pyrenees and the Massif Central act as mountain barriers; the "livas" in northern Greece and Bulgaria blows across the Balkan mountains; and the "leveche" is a hot southerly wind that blows across Spain. Beyond Europe, the "chinook" blows down the eastern flank of the Rockies and across the Great Plains of the USA and the Prairie Provinces of Canada, the "Santa Ana" is fiercely hot and desiccating wind that blasts across

Los Angeles and neighbouring coastal areas from the deserts of southern California, and the "berg" wind descends from the interior of South Africa to the coastal fringe of Cape Province and Namibia.

THE DAY THE MIDLANDS GROUND TO A HALT

Thousands of people will remember Saturday 8 December 1990 for the rest of their days. Most of them will have been travelling on that date, from London to Birmingham, or maybe to Sheffield, or perhaps Manchester. They may have planned a two or three-hour journey; they certainly would not have bargained for twelve or fifteen hours stuck on the motorway.

That snowstorm was variously described as "the heaviest since 1981", "the worst for a decade", or "the most severe for at least twenty years", according to your source of information. As far as one can measure the severity of snowstorms, none of these was true. Deeper snow, more severely drifted too, fell across southeast England in January 1987, a more prolonged snowstorm with widespread traffic chaos struck the Midlands in February 1985, while one of the worst blizzards of the century swept southern England, the West Country and south Wales in January 1982.

But an argument can be made for describing the Midlands snowfall that Saturday as the most dislocating for at least twenty years – and probably for very much longer. The chaos resulted from an exceptional combination of meteorological factors accompanied by an apparent disbelief by thousands of motorists that mere weather could disrupt their journeys. There were seven important factors that coincided to make this a most unusual snowstorm:

(1) The worst-hit regions had heavy rain before the snow set in, washing away salt and grit laid by highways departments in

response to severe weather warnings received the preceding night.

(2) In the Midlands the snow began to fall most heavily just before dawn on a Saturday morning, when motorway traffic levels are at one of their lowest points of the whole week. Steady traffic is one of the best methods of keeping roads clear of snow at temperatures near freezing.

(3) The snow had a high water content, making it wet, soft, sticky, and clinging. The water acts as a lubricant – wet snow is more slippery than ice.

(4) Wet snow does not normally drift, unless sustained wind speeds exceed 30 mph – unusual away from coasts and hills. But on this occasion sustained winds of 35 mph or more were recorded at Birmingham Airport, for example. Strongest gusts were in the region of 60 to 65 mph.

(5) The snow fell at a temperature close to or a fraction above zero – hence its wetness – but once it had stuck to power lines, windscreens, etc., it was subject to evaporative cooling, and the slush began to turn to ice. This has a terminal effect on overhead power lines and British Rail pantographs.

(6) The "plastering" effect due to wind and wetness resulted in sharply diminished visibility in motor vehicles, which could have led some drivers to slow sharply or to stop without warning. This was probably exacerbated by piles of accumulated slush and ice flying off high-sided vehicles at intervals during the snowstorm.

(7) The rate of snowfall is known to be highest at temperatures close to freezing point, and on that Saturday the rate of accumulation exceeded six centimetres per hour at times – very unusual over a large area in Britain. Once traffic stopped, fresh snow meant that within minutes it was incapable of restarting.

The newspapers had a field day. Having tasted blood after the Great Storm of October 1987, it was now normal practice for Fleet Street to demand that heads should roll after every weather disaster. On this occasion the forecasters were let off lightly, although forecasts of Saturday's weather made the previous day had not been terribly accurate. The scapegoats this time were the local authorities and the emergency services – who actually did an admirable job in the face of the impossible circumstances outlined above. On top of the appalling weather conditions, the snow-clearing crews also had to cope with the power cuts that affected motorway maintenance compounds. Police and ambulance services also suffered power cuts, and the Warwickshire police control room alone received over 30,000 emergency calls.

Angry MPs gave the transport minister, at that time Mr Malcolm Rifkind, quite a rough ride the following week at question time in the House of Commons. Most of the MPs' anger was just as synthetic and misplaced as Fleet Street's had been, but the Secretary of State promised a review of arrangements for dealing with severe weather. This was finally published in May 1991, and completely exonerated the emergency services and the various highways departments.

A NIGHT TO REMEMBER

Following a short broadcast I made on London's independent radio station, LBC, recounting the events of 8 December 1954 – when one of Britain's best-known tornadoes carved a path across west and northwest London – I received letters from a number of listeners who, 34 years later, still retained a very vivid memory of that night. The tornado – or maybe the first of a family of tornadoes – was first sighted on the Hampshire coast near Portsmouth at 3.30 pm, and the last authentic sighting was in east Hertfordshire about two hours later.

In the western and northern suburbs of London there were 12 known injuries, half of them at Gunnersbury tube station, which had part of its roof ripped off. A car was reported to have been lifted 15 feet into the air in Acton. Mr Owen Allen, who lived in Acton at the time, wrote:

> I noticed during the very cloudy afternoon lightning flashes in the distance – a bit late in the year, I thought. Quite suddenly, the whole world seemed to be filled with a deafening roaring wind. This fearsome noise lasted just for a very few minutes, then suddenly it was gone. Looking out into the street, there were chimney pots, slates, bricks and debris everywhere. The whole of Acton must be devastated, I thought. What of my wife and baby son, barely a mile away? I hastily grabbed the phone before anyone else did, expecting the worst. It seems, however, that this awesome experience was exclusive to a comparatively few of us because my wife knew nothing of it. It seems the whirlwind came from Chiswick way, and waltzed its narrow path through parts of Acton and Harlesden on its way to goodness knows where.

Miss Jessica Morton also remembered the day clearly, even though she did not experience the full force of the tornado:

> I have a lasting memory of the morning after this event. It was the year before I moved to my present address and I was then living in an upstairs maisonette on the south side of Belsize Grove off Haverstock Hill (southern fringes of Hampstead). From the upstairs back window of the attic bedroom I had a view of the surrounding roofs. The houses round there are almost all conversions, with bits stuck on and built on at all angles, and – possibly due in the main to war-time shortages – they were nearly all finished off with roofing felt. When I looked out of the window that morning all I could see was yards of loose, flapping, and missing roofing felt – not my own, luckily. I spent the rest of the day trying to match up the roofs with the front doors – no easy matter – and then knocking up the amazed occupants to alert them to the fact that they no longer had a weather-proof roof over their heads, since clearly they would be the last to know of it. I've forgotten long ago – as one does – lots of worse weather, but never THAT "morning after".

From Long Ditton, Mr R.F. Penfold sent in this memory:

> I am 66 years old and the year 1954 is remembered by me for two events. It was the year I got married, and it was the year of the tornado. At that time I was working in my father's estate agency at Willesden Green and I bicycled every day to work from Kingston-on-Thames – twelve miles each way. I left the office for home at about 5.30 pm on 8 December. It was a dirty night, raining and blowing, and I think there was also some thunder. I had my cycling cape on and cycled along Sidmouth Road, NW2, and then into All Souls Avenue, NW10. The latter road runs in a SW–NE direction and at the southern end there is quite a southerly slope and it is possible to see tall buildings at Isleworth as one travels in a south-westerly direction. I was half way down this incline when conditions deteriorated considerably, the wind increased, and there was a roaring sound. I quickly dismounted from my bicycle, got onto the pavement, and crouched by the side of the dwarf wall in front of a house, drawing my cycle cape over me. The roaring increased and slates and various objects flew about. It only lasted a short time, after which I continued on my way and very soon came upon evidence of the damage caused, with hoardings blown down across the road, etc. To this date I am still able to pick out scarred roofs in the line of the tornado, and at Willesden Green the wind ripped off some zinc sheeting from the roof of our office building and hurled it several hundred yards. I hope these recollections will be of interest. I know one's memory is inclined to play tricks after the passing of so many years, but I spent my whole working life in Willesden Green and did more-or-less the same journey for about forty years, so my memory was kept refreshed regarding what happened, and where, that night.

Chiswick and Acton lay in the path of the tornado during its most intense phase, so Helen Warth's story from Antrobus Road on the Chiswick–Acton border is particularly interesting:

> I was fourteen at the time and I remember the night of the tornado very well. We were sitting in our living room listening to the wind

when suddenly an extra strong gust came and all the windows made loud cracking sounds. At the same time we heard other crashes around the house. It was very frightening and we all rushed into the hall to get away. We were lucky and only suffered minor damage, even though a neighbour's chimney went through our roof. After the storm I walked round the streets and saw the damage others had suffered. It seemed to be houses and shops on corners which suffered most. One house I specially remember in Rothschild Road had the whole side blown down exposing all the rooms inside. I heard of a man in the same road who had just finished building up a special car (his pride and joy) and the wall fell on it. If the newspaper, the *Acton Gazette*, is still published and they keep back copies, you will find that they had a big coverage on this with pictures of Gunnersbury station minus the roof.

Clearly, a night to remember!

15. "Freak" weather

Freak weather is beloved of the news media, because it always makes a good story. But the word "freak" is grossly overused – any old thunderstorm gets described as a "freak electrical storm", a squall becomes a "freak whirlwind", and a few flakes falling in Wapping become a "freak snowstorm". When you think about it, weather phenomena are the last things that ever ought to be described as freaks. Dictionary definitions of the word usually refer to "abnormality" or use the expression "contrary to nature", but however unusual it is, weather is the last thing that should be considered to be abnormal or contrary to nature. Then again, you can't really imagine newspaper headlines reading "unusual thunderstorm" or "noteworthy snow shower", can you?

Weather can appear to be strange or newsworthy for a variety of reasons. I well remember that one of the oddest weather events I ever experienced was the first time I encountered thick fog in the Gulf (Arabian or Persian, depending on your allegiance). It may sound terribly mundane, but it was odd simply because it was so out of place. I had taken up a posting in Dubai in April 1978, and this was a late-July day in the same year. I had woken at 5.30 am, as was usual when I was on the early shift, and looked outside to discover a dense blue-grey fog shrouding the blocks of flats across the road. It was dawn, visibility appeared to be about 25 yards, and my bedroom was decidedly cool thanks to the air-conditioning. One's natural inclination was to shiver, since it so accurately recalled looking out at 8 o'clock on an English November morning.

So I was quite unprepared for the sensation that enveloped me when I stepped outside the front door. The temperature outside was somewhere near 33 or 34 °C (low 90s Fahrenheit), and the fog – infinitely more tangible than any English fog – hit me in the face like a warm wet flannel. The warmer the air is, of course, the more moisture it can hold, and within minutes water from the fog droplets was dribbling gently down my face. And my poor old body was also struggling manfully to cope with the exceptional humidity, so I was also sweating like a horse on this gloomy, grey, foggy early morning. Thankfully, the car also had air conditioning, so I could soon restore my own personal artificial environment, but here began another story – driving in thick fog in an Arab country. At the best of times it could be hair-raising, but at that time (things may have changed) the only concession the local drivers made to the appalling visibility was to switch on their puny hazard warning lights. Some even flashed their headlights at me because I had my headlights on; they thought I had left them on accidentally.

A weather event can also stick in the memory because it is something that we have not experienced before. During that first tour of duty in the Middle East I also encountered my first duststorm and its aftermath. Depending on the prevailing size of the particles of dust and sand that make up the local "soil", the wind will begin to lift them when it averages somewhere between 12 and 25 mph. The first effects are gentle swirls of dust, best observed as they blow across the road from the desert on one side to the desert on the other, rising no more than a few inches above the road surface. These are rather like the swirls of snow you get on occasion in Britain when the temperature is well below freezing and the snow is sufficiently dry and powdery.

On one such day following my early-morning shift I foolishly headed for the beach at around midday for a quick swim before returning home. The stiff easterly breeze blowing out of the interior meant that the usual northwesterly sea breeze had not set in around 11 o'clock as it usually did, and the heat was intense, probably approaching 45 °C (113 °F). While I was in the water, the desert wind and the sea-breeze effect suddenly combined forces, resulting in a 30 mph north-northeasterly wind, blowing straight down the coastline from the Strait of Hormuz. Towels and shirt vanished in the general direction of Abu Dhabi; mercifully shoes and socks (with car keys inside) remained. But the wind was now whipping up the sand to such an extent that it began to blot out the sun, and I was literally sandblasted as I raced across a couple of hundred yards of beach to regain the car park. Driving in the storm was not that memorable; it was a bit like a hot version of a snowstorm, with all the usual problems of poor visibility, a buffeting wind, obliterated road-markings, and incompetent driving by all concerned. Sand and dust are not as not as bad as ice, but dust is a surprisingly good lubricant, and steering and braking through dust drifts was quite exciting at times. What was memorable were the after-effects of this particular storm. When the wind drops, the larger particles soon drop out of the air under the influence of gravity, but tiny dust particles can remain suspended in the atmosphere for several hours, depending on how high up the material was carried while the wind was blowing. On this occasion, the wind dropped suddenly at about 3 o'clock in the afternoon, but a thick, choking dust haze remained for a couple of hours, too thick for the sun to penetrate, and combined with the continuing oven-like heat made this one of the most unpleasant afternoons I ever experienced while in Dubai. The suspended dust also rapidly clogged up the air-conditioning units, which had to be cleaned out at regular intervals, their loss of effectiveness adding to the general discomfort.

We may also remember otherwise uninteresting weather events because they occur out of season. I remember being told at the briefing before my first visit that it never rained in Dubai between roughly late April and mid-November, and the statistics seemed to bear that out. I consulted the Met Office's official record book, published in 1958, which showed that in 12 years of records at Sharjah (just a few miles from Dubai) there had been not a single day with rain in any month from May to October inclusive. I was also assured that clouds were a rarity between June and October. As the weeks passed I had no reason to doubt this, and the only interest in the sky was in the varying shades of blue and white according to how hazy the atmosphere was. But one day in mid-July there were some clouds. Not many, but they were quite pretty, and they actually became the main talking point amongst the Europeans and Americans. What did it mean? Would it rain? Had something gone dramatically wrong with the earth's weather machine? I was able to assure them, with the confidence provided by all that statistical evidence that, no, it wouldn't rain. Wrong! Early that evening it pelted down, for about five minutes. The gutters ran with water for a few moments, the sand-strewn streets turned into mud-slicked skating rinks, and another meteorological reputation hit the dust. It wasn't a freak. A careful analysis of the weather charts revealed that a shallow layer of moist monsoon air from the Indian Ocean had penetrated across Oman and into the southern portion of the Arabian Gulf. It was something that happened only at the peak of the Indian monsoon season, in late July and early August, and then only in a few years, but it had happened before, although clearly not during those 12 years of records that were quoted in the official reference book.

FREAK ELECTRICITY

Drop an intelligent Martian on planet earth, and the one aspect of the weather that might really surprise him or her would probably be the electrical activity. Studying our world from afar, the Martian would be able to deduce a great deal about the way the earth's atmosphere worked, and would know that the variety of weather phenomena depended on the composition and density of the air, on the amount and distribution of water both on the earth's surface and in vapour form in the air itself, and on the variation in temperature in both space and time. This would explain clouds and sunshine, rain and snow, and ice and fog. But our visitor would have to think a little bit more deeply before he came up with an explanation for thunderstorms.

We normally think that significant electrical activity in the atmosphere is confined to thunderstorms. It is certainly true that thunderstorms bring easily the most dramatic manifestations of atmospheric electricity, but it is equally true that there is always some electricity around, and in fine weather the ground is negatively charged and the air positively charged, and there is usually a very gradual seepage of electrical charge from one to the other. Under normal conditions we are unaware of this "static electricity", but when the lower atmosphere is very dry the rate of seepage is reduced, and this results in some very localized accumulations of electrical charge. These can be dissipated by human beings – for instance when hand touches car door, or more disconcertingly when fingertip touches fingertip, which can be quite shocking in more ways than one. Under ideal conditions you can even see the spark.

If nothing else were happening, the seepage would nullify the earth's electrical field fairly quickly. But thunderstorms put it all back again. The experts are now fairly confident that they know how thunderstorms work, at least in broad outline, although there are still some aspects of the electrical activity of these storms that are puzzling. Put simply, a positive electrical charge builds up at the top of the thunder cloud and a negative charge at the bottom, as a result of the powerful ascending and descending air currents and the rapid growth and shattering of raindrops and hailstones within the storm cloud.

When the difference in electrical charge between the top and bottom of the cloud (or between the cloud and the ground, or between the cloud and the air surrounding it) is sufficient, a discharge occurs, which we see as a lightning flash.

We all know the difference between forked lightning and sheet lightning. But the simple truth is that all lightning is forked, whether the discharge is wholly within the cloud, or between the cloud and the ground. However, we may not be able to see the discharge directly. Instead, we see a diffuse flash through cloud, or the hidden discharge illuminates other parts of the cloud or even other clouds. Sometimes at night, if the storm is many miles away, all we can see is a flickering of sheet lightning reflected on patches of high-level cloud or even on layers of haze. This is sometimes called summer lightning.

In the United Kingdom, the frequency of thunderstorms ranges from 4 or 5 days per year in the north and west of Scotland, to between 20 and 25 days per year in a few spots in the east Midlands and East Anglia. These statistics include observations of distant storms, and on only a few days will a thunderstorm pass overhead (say, within a mile) of a particular place, although on the other hand there are rare days when more than one overhead storm may occur. It is believed that in the most thunderstorm-prone parts of England approximately nine cloud-to-earth discharges occur per square mile, per year. Over the country as a whole, there are roughly 600,000 such discharges per year, and an average of five deaths by being struck by lightning per year. Thus, even in the middle of a thunderstorm there is only a one in 120,000 chance of anyone being killed by lightning.

In spite of these very long odds, there is a sizeable chunk of the population who have an intense dislike of thunderstorms – "brontophobes", you could call them. Some people even shut themselves in the cupboard under the stairs. And from our earliest years we are taught not to shelter under a tree or to touch metal window frames. But there is some good news if you are worried about being struck by lightning. A recent study, published in the Royal Meteorological Society's journal *Weather*, has shown that the risk of being killed by a lightning strike has diminished sharply over the last hundred years or so. In the second half of the 19th century there was a 1.6 million to one chance of a fatal injury in any one year; between 1920 and 1960 that had become a 4 million to one chance; and since 1960 it has decreased to a 13 million to one chance. Put another way, the number of fatalities from lightning strikes has dropped from over 20 per year to between four and five per year. The death rate is heavily tilted towards males, with men six times as likely to be killed by lightning as females.

A politician could doubtless weave wonderful stories around such statistics; however, there is a simple explanation for these rather startling figures. There has been no significant change in the frequency of thunderstorms in the UK during the last couple of centuries. But there has been a marked change in human activity during this period. As Dr Derek Elsom demonstrates in his article, the number of farmers and farm labourers has decreased dramatically, and these were usually the victims of lightning strikes earlier this century. Moreover, fewer women than men work outdoors. Important, too, is the fact that there is these days a greater awareness of the dangers of lightning and what to do if caught in a storm.

Thunderstorms are more frequent in the tropics than anywhere else on the earth, and nowhere more than on the Indonesian island of Java, where storms occur on an average of 220 days each year, although thunderstorms are observed on 322 days per year at Buitenzorg in the mountainous interior of the island. The climatologist C.E.P. Brooks estimated that there were some 44,000 thunderstorms every day on the planet, with roughly 100 lightning flashes per second.

FREAK OPTICAL EFFECTS

Professional meteorologists are not always the keenest observers of the skies and tend to be blasé about interesting and occasionally breathtaking optical effects that occur from time to time – assuming, that is, that they have even noticed them.

Perhaps they are slightly embarrassed to admit to the beauty so often displayed on the canvas that is the source of their daily bread. Keen skywatchers, by contrast, have an endless pageant above and around them to enjoy. . . and all for free.

Apart from vivid sunsets and sunrises, the only optical effect with which we are all familiar is the rainbow. But there are a host of others. And they are mostly due to the action of the sun's rays on water droplets, ice crystals, or dust particles.

Haloes around the sun (or the moon) and allied phenomena occur more frequently than rainbows, and can be quite stunning, but they are rarely noticed, although people very occasionally espy an arc of light broken down into the colours of the spectrum, and they sheepishly ask about "an upside-down rainbow with no rain in sight". Haloes and allied effects are the result of the interaction of sunlight and ice crystals in the upper atmosphere, and they are therefore associated with thin cirrus and cirrostratus clouds. But not all cirrus clouds cause haloes, and this perplexed early observers of the skies for many centuries.

We now know that the common halo, which subtends an angle of 22 degrees to the observer's eye, occurs when there is a predominance of icy crystals with faces at 60 degrees to each other – and this is probably the most common type of ice crystal. The rays of the sun are bent as they pass through the ice particles, a process known as "refraction", and the light is concentrated in a circle with the sun at its centre. The halo is frequently white, but the brighter it is, the more the prismatic colours are likely to be visible.

But ice crystals sometimes come in all sorts of shapes and sizes with all sorts of different angles between their surfaces, and these rarer forms result in the more infrequent sorts of halo phenomena. "Mock suns" are also very common, occurring adjacent to the halo on either side of the sun itself, and these are caused by elongated vertically aligned ice crystals. These mock suns (or "sun dogs", or more properly "parhelia") can vary widely in brightness, and are sometimes observed without the presence of a halo. They are frequently white, but can sometimes be brightly coloured, orange ones occurring particularly frequently. "Mock moons", or "paraselenae", are much rarer. Very occasionally part of the "parhelic circle" is visible. This is a ring of light parallel to the horizon which passes through the sun and therefore through any mock suns as well. Occasionally other mock suns are seen further around the parhelic circle; rarely one is found exactly opposite the real sun and this is known as the "counter-sun" or "anthelion".

One rarely seen optical phenomenon is the "green flash". It is a very transitory event – it lasts but a second or two – and is not often seen in Britain even by those who regularly look for it, because it requires a very clear, clean, haze-free atmosphere, and a sharp unobstructed horizon at sunset. These requirements are more commonly met at sea or in mountainous regions. The green flash occurs at the moment the setting sun finally disappears below the horizon. It is not a flash as such, but the last gleam of sunlight appears a bright emerald green. The phenomenon is due to refraction – the bending of light rays through different layers of air. When the sun is on the point of vanishing at sunset, the red and yellow parts of the spectrum no longer reach the observer, leaving just the green and the blue. The blue is absorbed by the atmosphere (that is why the sky appears blue), leaving the momentary green flash. In theory it is also possible to see this event at sunrise, but since it will appear at the first glimpse of the rising sun, you have to know precisely where you are looking.

Another interesting phenomenon is the "Brocken Spectre", so called because it can be observed fairly frequently from the summit of the German mountain called the Brocken. It occurs when a bank of mist or cloud lies just below the mountain top. If you stand with your back to the sun, you should be able to see your own shadow on the mist below. On many occasions the shadow of your head will be surrounded by circles of colour, usually quite faint, and this is called a "glory". The glory emphasizes the shadow, sometimes making it quite eerie. This particular phenomenon is now regularly seen by airline passengers – the shadow of the aircraft on a layer of cloud beneath is usually surrounded by such a glory.

THE NORTHERN LIGHTS

The "aurora borealis", otherwise known as the "northern lights" or the "polar lights", are a natural phenomenon that occur from time to time throughout polar regions of the world; in the southern hemisphere they are called the "aurora australis" or "southern lights". However, this beautiful phenomenon is also seen now and again in temperate latitudes. Very rarely, observations of the aurora have been made as far south in the northern hemisphere as the 40th parallel. These lights are electrical in origin, and take the form of very-high-altitude luminosity, sometimes very faint but occasionally brilliant, and which may be either multicoloured or virtually monochrome. They are caused by the injection of charged particles emitted by the sun interacting with the earth's magnetic field, causing a patterned electrical discharge, mainly between 100 and 300 miles above the earth's surface.

In practice, the aurora can provide a brilliant display of shimmering glowing colour, with blues, greens, yellows, reds, pinks, and mauves frequently represented. The brightest activity is usually in the polar half of the sky – the northern half in the northern hemisphere – and the displays may last only a few minutes, or may be observed through several successive nights.

These sometimes stunning light shows have been noted throughout history with even one or two biblical references, but it is only in recent decades that astronomers have linked these events with periods of unusually strong solar activity, when storms are raging in the sun's own atmosphere.

A major auroral display was seen on 9 March 1989 across practically the whole of the northern hemisphere as far south in Europe as the Mediterranean, and in North America as far as the Mexican border. This was probably the greatest such display since March 1946 or perhaps January 1938. Viewing such events in countries like Britain is very difficult because of the light pollution. Even outside the towns and cities there is so much seepage of artificial light outwards and upwards that any luminosity in the sky is diminished.

It is almost impossible to describe the astonishing beauty of a powerful aurora for someone who has never seen one. But this account of the March 1946 event as seen from Abernethy in Perthshire, published in 1947 in a slim volume entitled *Polar Lights* by Miss Cicely M. Botley, is a particularly good one:

> On the almost cloudless night of Saturday, March 23rd, there appeared about 10 o'clock, above the northern horizon, the bright glow so aptly named aurora borealis – the northern dawn. This is not an uncommon occurrence in the northern parts of these islands, and it excited little interest until an hour had passed, when the glow kindled rapidly into brilliance and, ascending slowly from the horizon, assumed the shape of a huge arc with the constellation Cassiopeia apparently resting on its summit. By contrast, the region between the horizon and the brilliant lower border of the arc appeared utterly black. It must be this region, now known as the dark segment, to which Aristotle refers in his "Meteorologica" as the abyss.
>
> Suddenly, the arc, which had remained perfectly quiet and regular during its ascent and development, broke into feverish activity along its whole length, dividing here into bundles of short rays, and there into diffuse pulsating patches of light. Tinges of red and green sparkled to enhance the yellow-white, which so far had been the prevailing colour of the display. The rays were leaping upwards, one bundle subsiding as an adjacent one darted ahead, and eventually the whole northern half of the sky was filled with streamers.
>
> The rays danced overhead into the southern half of the sky, apparently converging to the magnetic zenith – some 20 degrees south of the true zenith. The play of the rays in the corona round this point reveals that the magnetic field of the Earth exerts a directive influence on the agent responsible for the phenomenon.
>
> The coronal rays now merged to form what appeared to be a bluish-white vapour, sometimes like the smoke of the straw which is

burned in the country, at other times remarkably like cirrus clouds. Before long, however, the "vapour" resolved itself into rays, patches of green light waxed and waned in the east, and a quite fantastic cloud of brilliant red appeared and remained fixed in the western sky for almost an hour – so unreal that one imagined it to be a reflection of a gigantic fire below the horizon. It was such an illusion which caused fire engines to race towards the horizon in many parts of Southeast Europe during the great aurora of January 1938, and, much earlier, the soldiers of Tiberius to hasten to assist the inhabitants of Ostia, thought to be ablaze.

In the most intense phase of the display, waves of light surged up from the horizon to the zenith, fanning weaker rays and patches into brightness as they swept over them, like a breeze blowing over the dying embers of a large bonfire. By 2.30 am, the display subsided as suddenly as it had developed, leaving a weak, residual illumination in the sky, enlivened by slight sporadic ray activity.

ONCE IN A BLUE MOON

According to my *Collins English Dictionary*, "once in a blue moon" is an informal expression meaning "very rarely, almost never", and the *Longman Dictionary of English Idioms* concurs, suggesting "hardly at all or hardly ever".

Thus it could be said that blue moons occur rather more frequently than once in a blue moon, because occasional reports surface in the meteorological literature about once every ten or twenty years. The most famous example in Britain occurred on 26 September 1950, when both a blue moon and a blue sun were widely observed, while blue moons were observed virtually worldwide for several weeks after the Krakatoa volcanic explosion in 1883, and also more locally after an eruption of Cotopaxi in the Andes in 1880. A blue sun has also been reported through duststorms in Egypt.

The *Meteorological Glossary* defines the event thus: "A rare phenomenon in which more intense particle scattering of red light than of blue light makes the directly viewed luminary (i.e. the Sun or moon) appear blue or green. Application of scattering theory to the differential extinction of light measured in different parts of the visible spectrum indicated that the predominant radius of scattering particles (during the 1950 event) was about 0.5 micrometres [one 50,000th of an inch]." So now you know.

A certain Mr G. Bain Ross of Melrose in southern Scotland was the first to get his letter to the editor printed in the meteorological journals. He wrote to the editor of *Weather* magazine:

> I write to report a meteorological thrill. No natural phenomenon has ever caused such intense interest, speculation, and even alarm among a few people, as was produced in Scotland in the afternoon and evening of September 26th. The sun first began to assume a peculiar appearance shortly before four o'clock, when it was of a slate-grey colour, varying from that to a silvery shade. It was some time before it began to turn definitely blue, but by about a quarter past four the extraordinary change had taken place. Following the spectacle of the blue sun, the moon when it rose was observed also to be coloured blue. As the moon rose higher in the sky, however, the blue colour became less apparent.

Later, Colonel W.F.S. Casson of Crewkerne, Somerset, wrote:

> On September 26th I saw the moon rising and it was then a normal orange-yellow. At about 2300 BST I went out and noticed that the moon was now a most peculiar colour, something between a sea-green and a turquoise-blue. There were no stars visible, but visibility was quite good.

And Graeme Jackson, writing from Cambridge, had been in Birmingham that day:

> At 2300 BST on September 26th, from a point half a mile south of Selly Oak railway station, I noticed and remarked on the fact that the

> moon appeared distinctly blue in colour. This was all the more remarkable in view of the fact that the wind was in the north, bringing over industrial smoke and haze from the Black Country. Under such conditions the moon almost always appears reddish in colour.

Other remarks from official meteorological observers were received from Bwlchgwyn, near Wrexham, where the rising sun was seen to be of a bluish hue on the early morning of 26 September. This was the earliest report on this side of the Atlantic. The colour was described variously as steely grey, royal blue, deep blue, and even purple. The attention of some people was drawn to the event by the fact that a bluish sunlight was shining into the rooms where they were working. When the moon rose there were as many reports of green colours as of blue.

According to the *Meteorological Magazine*, Mr F. Bothwell of RAF Leuchars, in Fife, observed the phenomenon around mid-afternoon, and immediately arranged for an aircraft to investigate. The pilot, Flight Lieutenant West Jones, reported that the sun appeared blue to a height of at least six miles. At this level his aircraft entered a cloud of thick brown haze, which he did not clear until he had reached an altitude of eight and a half miles, by which time the sun appeared normal. The pilot of another aircraft entered the haze layer at approximately the same altitude above Cambridge the following morning, and reported a concurrent smell of burnt paper, and commercial pilots flying into and out of Prestwick, Ayrshire, noted similar events, and indeed similar smells. The discoloration of sun and moon was not confined to Britain, and several parallel observations were made in western Europe between the 26th and the 28th, notably in Scandinavia, Denmark, Germany, France, Switzerland, Portugal and Italy, and also at Gibraltar, but not in Vienna or in Malta, while in mid-Atlantic one British vessel reported seeing the sun displaying a distinctly mauve colour on 25 September.

It was easy to identify the cause of this peculiar event. Major forest fires had broken out on the eastern flank of the Canadian Rockies as early as 17 September. Alberta was the worst-hit province: the fires burned out of control for over a week, and were at their most extensive around 23 and 24 September. A huge pall of smoke developed during this period, and this was subsequently caught up in the jet stream at an altitude of five to eight miles above the ground, and was swept first eastwards and then southeastwards across Saskatchewan and Manitoba to southern Ontario, and thence across the eastern states of the USA. This thick smoky cloud was so dense that artificial light had to be used during the daytime in cities like Toronto and Buffalo, causing considerable concern and even some panic amongst the general populace, and this rather overwhelmed the excitement at the coloured sun and moon, which were also noted on the American side of the Atlantic. The pall of smoke gradually became more attenuated as it crossed the Atlantic, but it escaped being washed out by rain, and there was clearly sufficient material left in the upper atmosphere for the effect to be visible for at least 12 hours in any one place (cloud cover permitting) in Europe.

If you can't wait for the next once-in-a-blue-moon occurrence, you can make your own – simply by viewing the moon through bonfire smoke!

16. Holiday Weather

There was a time, not so very many years ago, when the climatic information found in holiday brochures was picked and chosen to provide the best possible gloss on the resort being advertised. This may still be the case, but it is almost ten years since I last picked up such a brochure in anger. A favourite trick would be to compare the resort with London, but this would not always be a straightforward, legitimate comparison. If the putative hotspot were actually not all that hot, then its average daytime maximum temperature for each month would be compared with London's overall mean monthly temperature – that is, the average of daytime and night-time temperatures. A typical instance might be the Azores, where the average July maximum temperature of 24 °C (75 °F) is not very different from London's 23 °C (73 °F), but it looks much more impressive if it is put side by side with London's overall July mean temperature of 18 °C (64 °F).

What to do about sunshine statistics in a cloudy resort? This is a more difficult problem. Again, take the Azores, where the port of Angra do Heroismo basks in an average of 6.3 hours of sunshine per day in July, on a par with London's 6.4 hours per day. You could, I suppose, ditch sunny London as your comparison point and use Manchester say, or better still Fort William. But no, that would invite derisive snorts. Much better to compare London's 6.4 hours of sunshine with the Azores' 14 hours of daylight. Sunshine . . . daylight . . . who is going to notice if you don't make a song and dance about it? It is, to borrow someone else's phrase, being economical with the truth, but you have not actually told an untruth.

In any case, climate statistics have to be used with great care, because they only divulge the broadest-brush information about what sort of weather you are likely to encounter on your holiday. They may tell you what the average temperature for a particular month is, but they do not tell you how often those average conditions actually exist. In some parts of the world there may be a close relationship between long-term climate and short-term weather, while in others there may be very little relationship at all.

The United Kingdom is a perfect example of a country where the basic climate statistics can be next to useless for any sort of planning, whether it be for holidays or for some commercial activity. There is such a wide range from year to year in any particular month, and in most years there are only one or two months that could remotely be described as "average", and then only over a limited portion of the country. One example illustrates this perfectly. At Plymouth, the highest temperature during the whole of July 1988 was 18.3 °C (65 °F), while in July 1989 the temperature exceeded that value on 28 days out of 31, with maxima averaging 23 °C (73 °F) and peaking at 29 °C (84 °F).

By contrast, if you look up the climate tables for Singapore you can rely on the figures being a good indication of the weather on most days during the year. They will tell you that the average daytime maximum temperature for the whole year is 30.8 °C (87.6 °F), and there is so little seasonal variation, and so little day-to-day variation, that the temperature will reach somewhere between 30 and 32 °C (86 to 90 °F) on at least 300 days during the year, and it may reach 34 °C (93 °F) on just one or two days per year, and 36 °C (97 °F) about once every thirty or forty years. London's mean annual maximum temperature is 11 °C (52 °F), but an afternoon high between 10 and 12 °C (50 to 54 °F) will occur on only fifty or sixty days during the year, the temperature may approach 30 °C (86 °F) on one or two days in an average year, and the forty-year extreme is close to 36 °C (97 °F) – the same as Singapore's.

You can prove anything with statistics. Did you know, for instance, that practically the whole population of Britain has an above-average number of legs? There is, as far as I know, not a single person in this country with three legs, but there are several with only one. So the average number of legs is fractionally below two – thus anybody sporting two such appendages has more than their fair share. This bizarre notion has its parallel in climate statistics. In desert regions, rain is of course rare, but when it falls, it falls in torrents. For instance, in Bahrain the average October rainfall for the period 1928 to 1974 was one-hundredth of an inch of rain, but this was entirely due to one single solitary thunderstorm when half an inch of rain fell in less than an hour. Thus 46 of those 47 Octobers were drier than the average.

Another point to bear in mind is that the normally available rainfall statistics do not tell you how hard the rain falls, how long it rains for, or whether it tends to rain at particular times of the day. Returning to Singapore, the average yearly rainfall is 95 inches, about four times as much as London, and significant quantities of rain fall on roughly half the days during the year, which is roughly 30 days more than London. But typical Singaporean rain is a sudden heavy downpour in the late afternoon or evening, perhaps lasting as little as half an hour, and rarely lasting more than four hours even on the wettest days. A typical British drizzle, however, can produce very small amounts in the rain gauge but may last over a period of several hours. Thus the nuisance-value of this sort of rain is very much higher than a short sharp shower in the tropics.

There is no way of knowing exactly what the weather is going to do on your holiday, but the message is, when you study the climate statistics, remember that they can only be the roughest guide to what you might end up with.

THE DORDOGNE

The Dordogne may be a second home to a large number of Britons, but they are probably unaware how similar the climate of the more hilly parts of the district is to that of southern England.

The chief town of Périgueux is 450 miles due south of London as the crow flies and has a population of 60,000 including suburbs, while the rest of the *département* is essentially rural. The Dordogne is as large as Norfolk and Suffolk together, and the land rises from near sea level in the lower parts of the valley of the River Dordogne, in the far southwest, to 1500 feet above in the far northeast, on the main road from Périgueux to Limoges. These rolling hills are the first high land encountered by moist west winds blowing in from the Bay of Biscay, and the land continues to rise eastwards in a series of ridges towards the western flank of the Massif Central.

It is in that northeastern sector where the similarities with, say, Hampshire are closest. In the small country town of Piégut, for instance, the average daytime maximum temperature ranges from 6 °C (43 °F) in January to 24 °C (75 °F) in July, and the average annual rainfall is 40.5 inches. This compares with 45 °F in January and 72 °F in July at Southampton, and 40 inches of rain per annum in the region of Alton and Petersfield. The most important difference concerns sunshine, with 1860 hours in the northeastern Dordogne compared with 1650 hours at Southampton, although the Isle of Wight resorts of Shanklin and Ventnor are very similar at 1800–1850 hours. Vineyards now speckle the south-facing slopes of the Hampshire Downs, but in this particular part of the Dordogne they are notable for their absence. This is probably as much to do with the poverty of the soils as with the climatic characteristics.

There is a marked change in the climate as one heads southwards and downwards, towards some of the most intensively cultivated wine-growing country in this part of France. At

Bergerac, for instance, average afternoon temperature rises from 9 °C (48 °F) in January to 26 °C (79 °F) in July, annual rainfall averages 34 inches, and sunshine 2150 hours. Curiously and coincidentally, these are similar to the values obtained in Hampshire and Sussex during the unprecedentedly warm and sunny year of 1990, when some of the best British vintages were obtained.

Extremes of weather are more extreme than they are in England. In August 1990 when the UK established a new all-time record of 37.1 °C (98.8 °F) at Cheltenham, the temperature in the southern Dordogne reached 40 °C (104 °F). This was not a departmental record, though, as 42 °C (108 °F) was recorded in July 1923. Even in winter, the weather can occasionally be warm and sunny, and 21 °C (70 °F) has been recorded once or twice in January, and 25 °C (77 °F) occurred in February as recently as 1990, while a heatwave in early October 1985 brought temperatures as high as 33 °C (91 °F). By contrast, in the northern Dordogne, readings below −18 °C (0 °F) have been observed on several occasions recently, especially during the winters of 1990–91, 1986–87 and 1984–85. During one particularly severe snap in January 1985 the temperature remained below −5 °C (23 °F) for twelve consecutive days, and on both the 8th and the 14th the daytime maximum temperature was −12 °C (10 °F). Overnight minima were as low as −21 °C (−6 °F) on the coldest nights during this spell.

Snow falls with about the same frequency as it does in southern England, very occasionally in November and April, and as one might expect it lasts much longer over the hills in the northeast compared with the lowlands in the southwest. But unlike southern England, when snow does arrive it is often in the form of a heavy fall of dry powdery snow, and it may last a week or more before suddenly vanishing.

The weather usually comes in more clearly defined spells than we are used to in England. During hot summers, drought and sunshine reaches Mediterranean proportions, which is hardly surprising as the *département* is only 200 miles from the Mediterranean coast. In June 1976, for example, no rain fell at all during the month, while the sun shone for just over 380 hours – an average of 12.7 hours per day, and afternoon temperatures averaged 26 °C (79 °F) in the north and 29 °C (84 °F) in the south. But even summer months can be dominated by winds from the Atlantic, and on such occasions the weather is just as relentlessly cool and dull and wet as it is during poor summer months in England. June 1992 was one such month, with six to eight inches of rain, 120–140 hours of bright sunshine, and afternoon maxima of 18–20 °C (64–68 °F).

Summer thunderstorms are notable for their severity, often originating over northern Spain, pepped up as they travel across the Pyrenees, and then given added impetus as cool moist air from the Bay of Biscay interacts with hot humid air from the Mediterranean. On some occasions, spectacular electrical activity lasting an hour or two may be accompanied by little or no rain. On others, the storm may be heralded by distant flickering lightning and low grumbling thunder, then as the storm cloud approaches, a dramatic squall of wind rushes noisily through the trees, bending them almost double, immediately followed by a violent cloudburst of rain and hail. Hail is the one thing that the vine-growers really fear during the summer season in the Dordogne, even more than drought, and in roughly one summer in three on average at least one prominent vineyard suffers seriously from extensive hail damage.

One peculiar characteristic of this region is the cold wet day that often follows the thundery end to a heatwave. While the storms continue to rage over the Massif Central, the Mediterranean air is displaced over the Dordogne by Atlantic air, which is in turn forced to rise over the ranges of hills, resulting in a steady downpour that may last 24 hours or more, sometimes depositing over two inches of rain. On such occasions, even in August, the afternoon temperature may remain near 11 or 12 °C (52–54 °F), perhaps following 32–35 °C (90–95 °F) the previous day.

SPAIN'S OTHER COSTAS

In October 1991 there was considerable embarrassment for the Spanish tourist authority when they were severely censured for claiming that visitors to the north coast of Spain could expect nothing but sunshine. This "Costa Atlantica" is completely unknown to the vast majority of the package-holidaying Brits that one finds strewn across the other Costas throughout the year, but clearly the Spanish Ministry of Tourism would like to encourage more visitors to that part of their wonderfully scenic country.

However, one look at the lush, verdant countryside pictured in the relevant holiday literature will indicate that incessant sunshine is not the norm in the top left-hand corner of the Spanish map. Indeed, even the Costa del Sol – the sunshine coast – can only claim a 67 per cent record for sunshine hours throughout the year, although this does rise to 80 per cent during the summer months. Along the Atlantic coast, the sunshine success rate drops below 40 per cent in some places.

Annual sunshine totals include 2040 hours at La Coruña in the extreme northwestern corner of Spain, 1705 hours at Gijon and 1767 at Santander, and 1795 hours at San Sebastian near the French border. The major city of Bilbao, just inland from the coast, averages 1720 hours per year, and a little further inland the town of Avala has a figure of 1680 hours, the lowest in the whole of Spain. La Coruña apart, these figures are all on a par with those recorded along the south coast of England.

If we study the figures of Gijon in a little more detail, we find that the annual total represents just 38 per cent of the theoretical maximum possible, and this hides a seasonal variation from 31 per cent in winter, 36 per cent in autumn, and 40 per cent in spring, to 43 per cent in summer. The cloudiest month of the whole year is December at 27 per cent, while the sunniest is August at just 48 per cent. Thus we can see that this part of Spain is rather sunnier than southern England in late autumn and winter, but it is significantly cloudier in high summer. Gijon's average monthly figure of 177 hours in June compares with 205 hours at Tiree in the Inner Hebrides and 175 hours at Stornoway in the Western Isles.

A look at the map reveals that Gijon and Santander are roughly 480 miles from the Costa del Sol as the crow flies, but they also face Land's End across a 480-mile stretch of the Atlantic Ocean – indeed the region's climate is largely controlled by its proximity to the Atlantic. It should therefore come as no great surprise that the north coast of Spain has a climate not very different from Cornwall's. Gijon's average sunshine total for the whole year of 1705 hours compares with 1752 hours at St Mary's in the Isles of Scilly, while the average annual rainfall total of 41 inches is exactly the same as Falmouth's. Meanwhile, July afternoon temperatures at Santander average 21.5 °C (71 °F), compared with 19.5 °C (67 °F) at Penzance.

So, if you find our West Country climate congenial during the summer, you should find northern Spain equally pleasant, but a second Costa del Sol it ain't.

Spain, of course, has another Atlantic coast – between Gibraltar and the Portuguese frontier – washed by the warm waters of the Gulf of Cadiz. The climate here could hardly be more different from the far north, with almost twice as much sunshine, making this the sunniest part of mainland Europe. At Cadiz itself, the sun shines for an average of 3243 hours per year, which translates to 8 hours and 53 minutes per day. Monthly values range from 175 hours in December and January to 390 hours in July. Put another way, this area gets 72 per cent of the theoretical maximum possible quantity of sunshine during an average year, and as much as 87 per cent during an average July.

Temperatures on this section of the coastline are moderated during the summer by onshore breezes, which provide a sort of natural air-conditioning during the hottest part of the day, and afternoon maxima during July and August average 27 or 28 °C (81–82 °F). But away from the coastline, across the plains of Andalucia, we find the hottest part of Europe, where summer afternoon temperatures average 36 to 38 °C (97 to 100 °F), and in the stifling cities of Seville and Cordoba most summers have frequent temperatures in excess of 40 °C (104 °F). It is therefore with some justification that this part of Spain is known as "the oven of Europe". The temperature has been known to reach 49 °C (120 °F) in Seville, but the highest value ever officially recorded in Europe appears to be 50.5 °C (123 °F) at Los Riodades in Portugal.

A third little-known costa, the Costa Almeria, lies in the southeastern corner of the country, just to the right of the Costa del Sol. This area, around the city of Almeria, was the last of the Mediterranean Costas to be developed for tourism, not because of any climatic defect but because of the difficulty of laying on a viable infrastructure, especially water supply, to support any sizeable resorts. Sunshine values at Almeria are almost as high as they are at Cadiz, averaging 3053 hours per year, and ranging from 175 hours in December to 365 hours in July. But this particular area is remarkable for its aridity, and exhibits the only true desert climate in the whole of Europe. Average yearly rainfall in the city is just 9.2 inches, and potential evaporation exceeds mean rainfall by a factor of four. The driest area of all lies just to the east of Almeria, extending past the Capo de Gata and as far north as the town of Aguilas, and in this district the average annual rainfall is less than eight inches. In a way, the Mediterranean holiday philosophy of wall-to-wall sunshine and as little rain as possible can be carried too far. Occasional thunder showers during the summer can be quite dramatic, but even if you are not excited by the prospect of thunder and lightning, the short sharp downpours will lay the dust, wash out the haze and pollution in the atmosphere, and freshen up the local plant life, which normally looks extremely tired and brown at this season.

DOWN UNDER I

New Zealand weather is a closed book to the majority of us in Britain, but very occasionally it does impinge on us. This is usually when important sporting events are under way Down Under, such as rugby or cricket Test Matches, and there was a rather longer period in 1990 when the Commonwealth Games took place in Auckland. And now that the first TV soaps based in New Zealand are appearing on daytime television in the UK, a rather larger audience may absorb just a little peripheral knowledge about the weather there.

School geography lessons teach us that Australia and New Zealand are "the Antipodes", a rather loose use of the word since it strictly means two points on the earth's surface *diametrically* opposed. In fact New Zealand is closer to the tropics than are the British Isles, and the southernmost city of Invercargill lies at a latitude of 46 degrees south, equivalent to central France, while the northern city of Auckland is situated on the 37th parallel, equivalent to the Costa del Sol.

Notwithstanding this, those old school textbooks also teach us that the climate is similar to Britain's. Any New Zealander will tell you that this is a gross generalization, pointing in particular to the fact that most parts of his or her country enjoy more sunshine hours than the United Kingdom does. It is certainly true that the country's weather is dominated by the belt of westerly winds that encircle the southern hemisphere in temperate latitudes (sometimes called the "Roaring Forties"), but those westerlies are much less subject to interruption by continental influences than their northern-hemisphere counterpart. That means that New Zealand's climate does not suffer the daily, monthly, or seasonal extremes that ours does, although the day-to-day weather is quite variable enough. And there are strong contrasts between the southern tip of South Island and the northern end of North Island, and also between east and west, especially on South Island where the Southern Alps act as a very efficient climatic barrier to those prevailing westerly winds.

Let us look more closely at those two cities at either end of New Zealand. At Invercargill, average afternoon maximum temperatures range from 9 °C (48 °F) in July to 19 °C (66 °F) in January and February. These figures are remarkably similar to, say, Plymouth, although night-time temperatures at Invercargill are several degrees lower. Rainfall averages 45 inches per annum, distributed evenly throughout the year, again pretty much the same as the coasts of Devon and Cornwall. Snow is not unknown, indeed snow and hail showers fall quite frequently in mid and late winter, but the snow rarely settles near sea level. As soon as you move inland, though, especially in the hills, one or two heavy falls of snow occur in most winters. Compared with London or Cambridge, however, the climate is cooler in summer, milder in winter, and considerably wetter at all seasons.

Auckland's climate is warmer than London's in every month of the year but especially in winter – frost is virtually unknown in the city, and snow is only seen on television. Rather contrarily, it is also wetter and sunnier than London in all four seasons. From this it can be deduced that when rain does fall, it falls more heavily and clears more quickly than it does in Britain, while those protracted spells of drab grey skies with nothing worse than a little drizzle that are so familiar to us over here simply do not occur over there. In January and February, the hottest months, afternoon temperatures typically reach 23 °C (73 °F) in Auckland (compared with 22 °C or 72 °F in July in Cambridge and 20 °C or 68 °F in Manchester), much lower than you might expect for the latitude, and this is due to the prevalence of stiff breezes from the ocean on most days when the sun shines. The temperature has been known to exceed 32 °C (90 °F) on average only once every 50 years. In the mid-winter month of July the mean maximum figure is 13 °C (56 °F), which compares with 7 °C (45 °F) during January in London and 6 °C (43 °F) in Manchester.

The average annual rainfall in Auckland is 49 inches, again distributed fairly evenly throughout the year, although there is slightly more in winter (5.7 inches in July for instance) than there is in summer (3.1 inches in January). The equivalent yearly figures for London and Manchester are 24 inches and 35 inches respectively. Auckland's January sunshine averages over 300 hours, or almost ten hours per day, and this is half as much as London's July figure, and almost double Manchester's.

DOWN UNDER II

Anyone following the Cricket World Cup tournament in Australia during February and March 1992 will undoubtedly have noticed how many matches were spoilt by rain. It is quite extraordinary to think that more games were interrupted by bad weather during that competition than in any of the three World Cups that took place in England – in 1975, 1979 and 1983.

The heavily populated eastern coastal strip of Australia has a reasonably wet climate, with Brisbane and Sydney both getting more rain in an average year than London, Manchester or Glasgow. The thing about Australian rain is that it falls more heavily and is therefore usually over quicker, allowing the sun to shine longer than it does here in the United Kingdom. Average annual rainfall is 45 inches in Brisbane, 46 inches in Sydney, 26 inches in Melbourne, 21 inches in Adelaide, and 35 inches in Perth.

Brisbane enjoys a subtropical climate, and the hot humid summers produce occasional torrential downpours, which have resulted in catastrophic floods in the city about once every ten years. Winters are drier and still pleasantly warm. Sydney has plentiful rain in all months, late summer and autumn being rather wetter than winter and spring. Summers are very warm and humid, while winter temperatures are akin to early October in England. Melbourne has about the same quantity of rain as London, distributed evenly throughout the year. Summer temperatures are similar to Sydney's, while winters are rather cooler, but – crucially – humidity levels are generally much lower in

Melbourne. Perth and Adelaide are both blessed with a Mediterranean-style climate – that is, summers are dry, hot and sunny, and winters are mild, changeable, showery, and occasionally very windy.

Clearly, then, Australia is not all beaches and barbies, but if you gave me the choice between Manly Beach and Bognor Regis I wouldn't have to mull it over for very long.

DALATANGI – JEWEL OF THE EAST?

Dalatangi sounds like one of those small Mediterranean resorts, on Sardinia perhaps, that you find in the glossy brochures, but which you may have difficulty locating in the atlas. So it would come as no particular surprise that it is occasionally the warmest spot in Europe. One such occasion happened in mid-January 1992, when the temperature at Dalatangi climbed to 18.8 °C (65.8 °F). The trouble is, Dalatangi is not in the Mediterranean. It is a desolate, windswept headland on the easternmost tip of Iceland, occupied only by a lighthouse and a weather station.

Needless to say, this particular January day broke the previous Icelandic national record for the month of 14 °C (57 °F) by several degrees, and it was also several degrees above the normal maximum value during high summer. The highest temperature occurred just before midnight on Monday 13 January, and it was accompanied by broken cloud and a strong westerly wind. The exceptional warmth was not confined to this one isolated spot: Akureyri, Iceland's third largest town, which is located on a sheltered fjord towards the north coast, recorded 17.5 °C (63.5 °F), while in Norway 13 °C (55 °F) was reported in the hills northwest of Oslo.

The cause of these highly abnormal temperatures was linked to the enormous high-pressure system that was anchored over the British Isles for much of the winter, and which kept the weather remarkably dry throughout the season in the UK. Winds over the Atlantic were diverted northwards around the periphery of the high, flooding Iceland with warm moist air of subtropical origin. On arrival, this air stream was forced to climb over the mountainous interior, depositing copious quantities of rain and snow, before reaching the far side of the island as a dry wind that was warmed rapidly as it descended towards the coastline – a dramatic instance of the so-called "föhn effect" (see Chapter Three).

EXOTIC WINDS

Strangely, for one of the windier countries of the world, Britain has comparatively few names for localized winds of different directions and different characteristics. The only one recognized by the *Meteorological Glossary* is the "Helm Wind", which is a cold gusty northeasterly wind blowing across the northern Pennines in the neighbourhood of the Cross Fell ridge into the Eden valley in Cumbria. It takes its name from the near-stationary wave-cloud formation that sits above Cross Fell and is typically shaped like a hat or a helmet.

That is not to say that our forebears ignored the influences of the wind on their lives. Far from it: Richard Inwards, in his collection of ancient country weather lore, found room for twenty pages of sayings and rhymes referring to the wind. But the only winds that he names are foreign ones.

In marked contrast to the British Isles, the Mediterranean region has a plethora of wind names. Arguably the most familiar to us is the French "mistral", a violent northerly wind that is funnelled along the Rhône valley and disgorges across the Mediterranean coast between Marseilles and Nîmes. In winter it is frequently intensely cold and dry, causing excessive wind-chill, and sustained windspeeds over 80 mph have been measured. The mistral is actually one of a family of winds, all of polar origin, seeking access to the Mediterranean basin through a variety of gaps in the European mountain ranges. It becomes the "maestrale" at Genoa, the "maestro" over Sardinia, the "bora" at Venice and Trieste, the "bise" in Switzerland, and the "vardarac" at Thessaloniki in northern Greece. Other cold winds from a northeasterly quarter, although generally less fierce than the mistral and its cousins, include the "gregale", which affects Malta and Sicily; the "tramontana", which is found along the west coast of Italy, including the Rome and Naples regions; and the "levanter", which is a moist easterly wind notable along the Costa del Sol and at Gibralatar.

Southerly winds from the Sahara are hot, dry and dusty when they first reach the Mediterranean basin, but they pick up considerable moisture by the time they reach continental Europe, thus becoming humid and thundery. The best known are the "scirocco", which is best observed in spring over Malta, Sicily, and southern Italy, and the "khamsin" over Egypt. More local names include the "leveche" in southern Spain, the "ghibli" in Libya, the "chili" in Tunisia, and the "simoom" in parts of Egypt. Related to these are the "föhn" (discussed in an earlier chapter), which is sometimes known as the "schneefresser" or "snow-swallower" in Austria and Switzerland, the "gharbi" in Greece, the "marin" on the southern French coast, and the "autan" in Aquitaine and the Périgord. The 19th-century German writer T. Fischer described the experience of a Sicilian scirocco thus:

> ... the air is misty, the sky yellowish to leaden, filled with heavy vapours, through which the sun can be seen only as a pale disc if at all. Man feels languid and oppressed, and disinclined for mental activity, and animals also suffer from these hot dry winds. Everyone stays at home as much as possible and does nothing. When the scirocco is especially hot, its scorching breath does great injury to the vegetation; the leaves of the trees curl up and fall off in a few days, and if it sets in when the olive trees and vines are in blossom a whole year's harvest may be lost.

Summer winds are particularly notable in and around Greece, where they may blow for days on end under a deep blue sky. These include the "étesian winds" in the Aegean Sea, and the "livas", which blows down from the mountains and across the narrow coastal plain of northern Greece.

The north Europeans appear to be as prosaic as ourselves, preferring in the main to call winds by their directions, but one small exception is worth noting. Norwegian fishermen refer to the "sunset wind", which blows on summer afternoons and evenings from between north and northwest – the direction in which the sun sets at that season – around the coastline of southern and western Norway, reaching its strongest mid- to late-afternoon, dying away again after the sun goes down. This is considered to be a sign of settled weather.

17. Sporting Weather

You would have thought that with a climate such as ours, we would have invented some summer sports that were less at the mercy of the weather than tennis, golf, and cricket. Over the years, Wimbledon has experienced some extraordinary meteorological interventions, golfers and cricketers have been struck by lightning, and at different times spectactors have been treated for heatstroke and for hypothermia. Cricket, at intervals during its history, has sought to allow the vagaries of the weather to influence the course of matches, providing an added dimension of uncertainty for the players to exploit. Two of our premier spring-time events, the Derby and the Cup Final, have taken place in blistering heat one year, and in arctic winds and snow showers the next. Meanwhile the winter sporting programme, although largely immune from the effects of rain and wind, has been annihilated during severe weather when ice and snow and frost (fog, too) inevitably come out winners. In the famous London smog of 1952, several Saturday afternoon football matches were cancelled because the capital's transport system was so chaotic that there was little probability of players and officials reaching the grounds, let alone any spectators.

Britain is, of course, not alone in having its sporting activities seriously disturbed by the weather. First-class cricket matches have been abandoned without a ball being bowled in all the major cricket-playing countries from time to time, and major golf tournaments in the USA have had to be suspended not only for torrential rain, but also while thunderstorms raged overhead, for sudden hailstorms, and even because a tornado-watch was in operation in the district. In some parts of Europe where snow and ice are regular visitors during the winter, the soccer season is divided into two separate parts, one lasting from August or September until early December, the other beginning during March.

In the wealthy oil-producing states of the Arabian Gulf area, most of the sheikhdoms boast one tournament-quality golf course with proper greens and grass-covered rough, but the expense of the daily watering required to keep these sporting shop windows in prime condition must be collosal. Other golf courses in these countries make use of the local environment: greens are replaced by "browns", which are regularly oiled and rolled to maintain as smooth and hard a surface as possible, while the various degrees of rough are provided by sand of varying unevenness and softness.

In Bahrain, the local cricket season is necessarily a compromise between the months of rather lower humidity and the length of daylight. In November, December and January, when weather conditions are at their most pleasant, the day-length is rather too short for a worthwhile match. But, especially in September, summer humidity levels are excessive. The United Arab Emirates now has a fully fledged cricket stadium at Sharjah with a field laid with imported turf, but I recall playing a couple of matches in September 1979 at the Awali Cricket Club in Bahrain, where the outfield was hard sand, and the pitch itself consisted of matting laid on top of concrete. The afternoon temperature in September hovers around 35–38 °C (95–100 °F), with relative humidity between 70 and 80 per cent, and in such conditions the perspiration pours off you even without any physical exertion. We had drinks intervals every thirty minutes when it was suggested everyone should consume at least half a pint of water. Even so, I still suffered quite serious heat exhaustion at the end of the day.

In Middle Eastern countries, the equivalent of the unexpected

British snowstorm is the sandstorm. Occasional games of cricket in Bahrain or Dubai have to be cancelled for this reason, and important soccer matches in other countries – especially in Egypt – have been abandoned or postponed when severe sandstorms or duststorms have swept in. The same has also been known to happen in Argentina.

Notwithstanding these peculiar problems faced by sportsmen overseas, some of the stranger interventions have occurred in Britain. It is cricket that appears to suffer most, partly because of its peculiar sensitivity to adverse environmental conditions, and partly because of the length of the game. Snow has seriously affected some county matches, not only at the very start of the season, in April, but memorably in 1975 in early June, and in April 1981 the combination of strong winds and severe cold produced such a large chill-factor that the umpires took the players off the field as they considered the conditions were "dangerous" (see Chapter Six). Strong winds are high in nuisance value, not just because of debris blowing across the ground, but because an average wind speed of 20 mph or more will blow the bails off the wickets – in these circumstances special heavy bails are used, but even these are virtually useless if the sustained wind reaches 30 mph. County Championship matches have been suspended due to excessively bright sunlight, either reflecting from nearby windows or from car windscreens, or simply shining in the batsman's eyes when it is very low in the sky.

There are, of course, occasional days when sporting activity can proceed without a thought for the weather, and with a bit of luck the activity on the ground or the field or the course may be quite interesting as well.

TENNIS

The most enduring meteorological memory of Wimbledon during the last fifty years must surely be the sun-baked Championships of 1976. The grass turned brown, the courts became arid and dusty, spectators dropped like flies on the hottest afternoons, and it was no small wonder that the players themselves did not suffer more from the heat. For the 1976 Wimbledon fortnight coincided almost precisely with the hottest fortnight of that record-breaking summer – indeed, there was no question that, in the London area, this was the hottest fortnight since at least 1783 and probably for much longer.

Official weather records are available for Kew Observatory – just five miles west of Wimbledon – and also for Heathrow Airport, which is six miles further to the west. The weekend that preceded the tennis was hardly auspicious. It rained all day Saturday (there was no play at all in the Lord's Test Match on that day), and the Sunday brought a few showers and a blustery breeze. But by the first Monday – 21 June – pressure was rising across the country, the weather was settling down, a few spots of rain fell between one and two o'clock in the afternoon but otherwise it was a dry day, and in spite of rather restricted sunny spells the temperature climbed to 23 °C (73 °F). Tuesday the 22nd brought a Mediterranean blue sky, virtually uninterrupted sunshine, and an afternoon high of 27 °C (81 °F), and from then onwards there was no change right through until the final Saturday. The temperature soared into the 90s Fahrenheit (above 32 °C) on 25, 26, 27 and 28 June, and again on 2 and 3 July, and hottest of all were Saturday the 26th, Sunday the 27th, and Saturday the 3rd, each of which reached 34 °C (93 °F) at Kew. On 26 June, unofficial reports from reliable amateur weather stations included several figures near 36 °C (97 °F) in the western and southern suburbs of London, but these amateur stations are generally located in rather enclosed suburban gardens and are thus subject to a certain amount of overheating in comparison with relatively open sites like Kew Observatory. Over the thirteen days of the Championships, the average afternoon maximum temperature was 30.7 °C (87.3 °F) at Kew and 31.4 °C (88.5 °F) at Heathrow, and sunshine averaged 12.5 hours at Kew and 12.6 hours at Heathrow. Thunderstorms finally broke out over Wimbledon during the early hours of the Sunday morning following the Finals.

Now, all these official figures were at odds with the incredible figures published in the newspapers, and talked about incessantly by the commentators on radio and television; the temperature on centre court, it was said, reached 116 °F. Similar apparent discrepancies arise every year during Wimbledon when hot weather prevails. In 1994, for instance, tabloid front pages were decorated with a brightly coloured "108 degrees" amid the usual "Phew, what a scorcher" clichés, when the shade temperature in London actually stood at 29 °C (84 °F). The reason is simple: if you put a silly thermometer in a silly place, you get a silly temperature. If a thermometer is in direct sunlight it will simply absorb energy from the sun, and the mercury will measure the temperature of the glass out of which the thermometer is made. It will not measure the temperature of the air. Similarly, if the thermometer is ostensibly in the shade but resting against an object – a heavy roller, say – which is in the sun, the thermomoeter will measure the temperature of that object. Different materials and different colours absorb solar heat at different rates. The roof of a black car will feel much hotter than the roof of the white car next to it, even though both are equally exposed to the sun's heat. Grass, for instance, feels much cooler than sand or tarmac. And human beings feel somewhat cooler in light-coloured clothing compared with dark clothes.

Oh, and who won? None other than the Ice Man and the Ice Maiden – Bjorn Borg and Chris Evert.

The 1976 fortnight was one of a very small handful of Wimbledon Championships that have suffered no signficant interruption from the weather. Curiously, 1977 (Bjorn Borg and

Virginia Wade) was also virtually rain-free, although it was very much cooler, but the next dry one did not occur until 1993 (Pete Sampras and Steffi Graf). Going back rather further, Wimbledon 1957 (Lew Hoad and Althea Gibson) was almost as hot as 1976, with the temperature in north London reaching 35.6 °C (96 °F) on 29 June.

During the 1980s, however, rain was a much more frequent visitor to Wimbledon than drought and heat, and in several years the tournament authorities had their work cut out trying to fit in all the matches so that the Finals would take place on the scheduled days. They failed in 1988 thanks to a final Sunday of incessant rain, and only just succeeded in the very wet years of 1980, 1985, 1990 and 1991. In 1987, the first day was completely washed out, but the weather relented after the first five days, and the rest of the fortnight was warm and sunny. Some of the lowest temperatures ever recorded during the Championships have also occurred in relatively recent years, notably on Tuesday 1 July 1980, when the mid-afternoon temperature was just 12 °C (54 °F).

Some myths have arisen over the years, partly as a result of a sort of Chinese whispers – stories get passed from journalist to writer to broadcaster and back again with unintentional distortions and elaborations, and nobody bothers to check them out. Thus we are led to believe that the 1923 Championships were ruined because it rained on every single day, whereas the truth is that only one day brought significant rain that year. The really wet year actually occurred in 1922, although there were occasional dry days. Wednesday 5 July was a total washout, and further heavy rain fell on the subsequent Friday and Saturday, resulting in a huge backlog of matches, which meant that the tournament was not completed until 12 July – the following Wednesday.

Another rather silly elaboration concerns one of the first beneficial uses of the rainfall radar system in 1985. It was Men's Semifinal day – Friday 5 July – and after an unsettled and often rainy fortnight, the weather appeared to be playing ball at last, with better weather predicted for the weekend. One commentator has written:

> The Centre Court had a narrow escape, thanks to the quick thinking of the Met Office chief forecaster. Watching the rainfall radar he noticed a thunderstorm brewing over southwest London. A quick phone call to the tournament referee led to play being halted, despite the cloudless skies overhead. Fifteen minutes later the heavens opened, and after a short and sudden downpour, the rest of the day was clear.

The truth is that that Friday was a warm, humid, cloudy day, with thunder expected. Certainly the telephone call happened, but the storms had been tracked over Hampshire and Surrey for some time, and play was not halted. Rather, the start of the first of the semifinals was delayed until the storm had passed, thus obviating the likelihood of having to take the competitors off after just ten minutes or so of play. The sky was not clear overhead; it had been a cloudy morning and the storm clouds were already massing in the southwestern sky, and the weather remained largely cloudy for some time after the storm itself had moved away. The Centre Court was not "saved"; without the radar information, play would have been suspended the moment the rain started to fall, and the covers would have been put in place with the usual Wimbledon efficiency in a matter of seconds.

GOLF

Golfers are curious animals. They carry on playing in thunderstorms although they know the dangers of being struck by lightning. This may be because most players who have been struck survive. Or it may be that they accept the fallacious logic that because someone has been struck on their golf course in the past, no one will be again – the "lightning never strikes twice in the same place" syndrome. Oh, but it does. . . Or it may simply be that their game of golf is more important than worrying about a bit of adverse weather. To anyone with any sort of sporting passion, this latter reason rings truest.

Ironically, one lightning victim who survived is David Brooks, a talented golfer, and for many years the weather-man on Anglia Television. A slightly more famous golfer, Lee Trevino, also survived a lightning strike, along with his two playing companions, on 27 June 1975. It happened at the Butler National Golf Course at Oak Brook, Illinois, and the three golfers were sitting on the ground sheltering under umbrellas. They were all stunned and taken to hospital, where they were kept in overnight for observation. Trevino's back was burned, though not seriously, otherwise no permanent injury occurred. Disastrous lightning strikes occurred in the USA in September 1937 when four golfers were killed and three others injured at Pittsburgh, and in July 1941 when three caddies were struck and killed at Louisville, Kentucky. Lightning fatalities on British golf courses happened in May 1976, near Ipswich, and in August 1973, near Birmingham.

The premier event of the British golfing calendar is called, with just a sneaking hint of pomposity, not "The British Open Championship", but simply "The Open", and takes place over four days during the middle of July. Typical British seaside links golf courses even in high summer will usually produce a mixed bag of weather, with some sunshine, some showers, and occasionally blustery winds, and these contribute to the uncertainty factor. In one recent year, however, the weather was so bad throughout the four days that for competitors and spectators alike it turned into a pretty fierce endurance test. It was the 1987 Open, at Muirfield, near Edinburgh, and it took place between Thursday 16 July and Sunday 19 July. Afterwards, the golfing correspondent of London's independent speech radio station, LBC, described conditions to me as "the most relentlessly awful I have ever experienced at a major golf championship anywhere in the world. I was glad when it was all over."

The first half of July 1987 had brought plenty of good

weather with warm sunshine on most days – long overdue following a very poor June. But the weather began to change again on 14 and 15 July, as the high pressure of the preceding fortnight retreated into northern Scandinavia, and a deep depression near Iceland began to migrate towards the British Isles. On the first day of the championship a complex low-pressure area covered the UK with the main centre just off Northern Ireland, while an active cold front moved slowly near the east coasts of Scotland and England. Over the next three days, the principal centre of low pressure described a circle as it first slipped southwards into the Southwest Approaches, then turned northeastwards across southern and southeast England on the Friday, headed north into the central North Sea and then west towards Northumberland on the Saturday, finally veering southwest and then south across England and Wales during the Sunday. Rain fell for long periods on the first three days, and on the third day was accompanied by a gale-force northeasterly wind and a mid-afternoon temperature of 12 °C (54 °F). The final day brought a marginal improvement, but at least there was one bright spot for the home crowd – Nick Faldo won.

Excessive heat is rarely a problem during The Open, although in 1989 at Troon on the Firth of Clyde, of all places, all four days were very warm and humid. But in the USA, where many of the leading golf courses are inland, extreme heat is a frequent hazard. The highest temperature recorded during the US Open appears to have been 37 °C (99 °F) at Minneapolis in 1930 and Dallas in 1952. But the prize for the most unpleasantly humid weather must go to 1964 when the venue was Bethesda, Maryland – just outside Washington DC. Maximum temperatures of 32 °C (90 °F) and 35 °C (95 °F) were recorded on the last two days, accompanied by relative humidity between 70 and 80 per cent. It was just as hot and only fractionally less humid during the 1909 US Open at Englewood, New Jersey. The American weather historian David Ludlum relates the following story:

When asked about the heat at Englewood, Tom Vardon, a brother of the British champion, said: "After I had played about six holes in the morning, I felt certain I would never finish. I've never experienced such heat before and I don't want to again."

"Why don't you take off your coat?" a reporter asked.

"Take off my coat! I wouldn't think of doing such a thing on the other side."

Next day, however, the reporter had this to say about Vardon: "He dropped his coat yesterday and charmed everyone by a pea green shirt, lighter in tone than his green felt Alpine hat."

CRICKET

The 1250th Test Match was played during 1994, and the history of international competitive cricket extends back to March 1877 when Australia beat England at Melbourne in the first ever Test, although it has to be said that it wasn't called a Test Match at the time, and the two teams were not called England and Australia. For half of that history, Test Matches were normally completed in three days, but during the last forty years or so they have usually extended over five, and occasionally six, days. A small number of matches had no time limit – the so-called Timeless Test Matches. The longest of all involved England and South Africa at Durban in March 1939: it spanned twelve days including two rest days and one day abandoned because of rain, and even then was left drawn, as the English tourists had to return to Cape Town to catch their ship home.

Of those 1250-odd matches, only four were completely lost to the weather. The first two of these were England v. Australia Tests at Old Trafford, Manchester, in 1890 and 1938, and this merely confirmed the unfair reputation that Manchester has for wet weather, a reputation which is therefore particularly strongly held amongst cricketers and cricket-followers. Another England–

Australia match was abandoned on the 1970-71 tour, scheduled to run from 31 December to 5 January at Melbourne. And finally, England's match against the West Indies at the Bourda Oval, Georgetown, Guyana, was lost to the rain in March 1990. There were other near-misses, notably at Lord's Cricket Ground in London in June 1902 when only 105 minutes' play was possible, and at Trent Bridge, Nottingham, in June 1926 when play lasted just 50 minutes. The cricket ground at Georgetown is particularly prone to flooding as it lies almost exactly at sea level, just a stone's throw from the sea wall, so drainage is a serious problem, and the Guyanese capital has an annual rainfall over twice that of Manchester.

The domestic cricket season is often seriously disrupted by rain, especially during late April and May. If the early-spring has been wet, grounds have had little opportunity to dry out, and any further rain once the season has started will lead quickly to waterlogging. During the notoriously wet May of 1979, 34 out of the 37 county championship matches failed to reach a conclusion, and altogether ten matches were abandoned without a single ball being bowled. May 1967 was almost as bad. Meanwhile, the effects of extreme cold and even snow have been described in earlier chapters. Cricketers are just as much at risk from lightning as golfers, but there is no record of a professional cricketer being struck on the cricket field. It is a different matter, though, as far as club cricketers are concerned, and strikes to ground on cricket fields (when in use) seem to happen about once every other year. The last cricketing fatality was at Isleworth, southwest London, in 1992. On an earlier occasion, an umpire was hit by lightning and survived, but, so the story goes, the metal knee joint he had had following a serious accident in his youth was welded solid.

The cricketing authorities have had varying attitudes over the years towards the extent to which rain should be allowed to affect the playing surface – that is, the wicket itself, rather than the outfield. Over the last twenty years or so, Test Match wickets have always been protected from rain, although prolonged covering of the wicket during wet weather can lead to the wicket "sweating", and this additional moisture sometimes helps the bowlers. However, rain-affected wickets do not always provide additional assistance to the bowling side, and a dusty cracked wicket giving assistance to spin bowlers may actually be subdued by rain, while heavy rain on an already damp surface may deaden it completely – a "pudding", in cricketers' parlance. But it is probably fair to say that in the majority of cases the bowlers do benefit. In particular, if a heavy downpour is followed by hot sunshine, a superficial crust will soon form and this should allow the spinning ball to grip, and thus to turn sharply. This is a "sticky wicket", or, in Australia, a "sticky dog". Classic sticky-wicket conditions occurred in the first ever Test Match between Australia and India at Brisbane in late 1947. Australia amassed 318 for three wickets on a good wicket, but following heavy rain, the next five wickets fell for 62 runs before Australia declared, and India were subsequently bowled out for 58 in their first innings and 98 in their second.

The average annual rainfall at English Test Match grounds ranges from 24 inches at the Oval to 35 inches at Old Trafford. But some overseas grounds have much higher rainfall, although it has to be said that British rain tends to last longer while other countries tend to have sudden sharp downpours that soon clear away. In Australia, Brisbane averages 45 inches of rain per year, heaviest during the cricket season, and Sydney gets 47 inches, while Adelaide is driest with only 21 inches. In New Zealand, Auckland's annual rainfall is 49 inches and Wellington's 47 inches, but Christchurch is much drier with only 25 inches in an average year. South Africa's wettest venue is Durban with 40 inches, while the driest is Cape Town with 21 inches. In the West Indies, Georgetown's 89 inches contrasts with just 31 inches at Kingston, Jamaica. Meanwhile in the Indian subcontinent, Karachi is the driest of all the regular Test Match grounds, with a normal yearly rain of just 8 inches, and Colombo (Sri Lanka) the wettest with an annual average of 93 inches.

Curiously, Test Matches in England have largely avoided our most outstanding heatwaves, and no international cricket at all was played during the hot summers of 1881, 1900, 1901, 1906, 1911 and 1923, and there was only one match in 1932. Arguably, the hottest Test Match in England was the Lord's Test of 1975 between England and Australia where the temperature reached 34 °C (93 °F) on the fourth day, and the great heat encouraged Test Cricket's first streaker. In complete contrast, the England v. New Zealand match at Edgbaston, Birmingham, in 1965 was conducted almost throughout in a biting northeasterly wind and under depressingly overcast skies. The second day (28 May) was the coldest, with a midday temperature of just 8 °C (46 °F), players wore three or four sweaters, and hot beverages were supplied during the drinks intervals. It was very nearly as cold on the second day (7 June) of the England v. West Indies game at Headingley in 1991, where the mid-afternoon temperature stood at 9.5 °C (49 °F). What visiting West Indian umpire Steve Bucknor would have made of such weather one shudders to think – he was moved to wear gloves during the first match in England in which he officiated, at Trent Bridge in 1994, when afternoon temperatures were comparatively tropical at 14–16 °C (57–61 °F). Those two occasions are easily the coldest on which Test cricket was played anywhere in the world; the hottest venue is undoubtedly Adelaide where the temperature has been known to reach 49 °C (120 °F) during the cricket season, although it is not known which was the hottest Test Match day.

DERBY DAY

It was among the 20th century's worst sporting weather disasters. The Derby Day storm of 31 May 1911 was one of the most violent electrical storms ever to hit London and its suburbs, and in the metropolitan area 17 people lost their lives. There were actually several distinct storms that afternoon, but the worst of them appears to have been broken over Epsom Downs towards the end of the day's racing, and at nearby Banstead two separate rain gauges measured respectively 3.59 inches and 3.54 inches. A little further away, Chipstead reported 3.00 inches and Bletchingley 2.97 inches. Meanwhile in Epsom town, according to the official observer, 2.86 inches of rain fell in just over two and a half hours, most of which fell in just fifty minutes. The electrical activity was stupendous. The well-known amateur meteorologist Spencer C. Russell, who lived at Epsom, wrote:

> Fork lightning was first seen at 4.59 pm, and from then until 7 pm lightning and thunder were practically continuous. Rolling thunder was entirely absent, the peals coming in sharp decisive cracks, closely resembling cannonading, whilst the lightning flashes were of dazzling intensity. Between 5.30 and 5.45 pm a count of flash frequency yielded 159.

Russell also determined that, within a three-mile radius of Epsom, three people were killed instantly, 14 more were injured, one of whom died later, four horses were killed, and three hay ricks were set on fire. It emerged that none of the deceased horses was a race horse. Of the three people who died, two youths who had been sheltering against a reservoir wall were struck, while the other was a man driving a horse and trap across the Downs.

Russell has a worthy successor as Epsom-based amateur meteorologist in John Bird, who compiled a history of Derby Day weather to mark the bicentenary of the race-meeting on the Downs above the town in 1979. He quotes the editorial from one of the local newspapers:

> It would have taxed the skill of the finest word-painter to describe the scene at the height of the storm. It was an inferno of water, mud, thunder, lightning, and hail. Innumerable cars *hors de combat*, horses plunging with fright, a confusing heap of figures inextricably jumbled together in narrow roadways, half-drowned pedestrians, drenched cyclists, terrified women and children, and battalions of

> men helpless against the mighty powers of nature in one of her savage moods.

The railways in the district were also in serious trouble, with landslips at Merstham and Coulsdon, and flooding to the height of the train-footguards at Epsom.

The bicentennial meeting in 1979 almost saw a repeat performance. The whole of the week was unsettled and thundery, and on 31 May (again) a terrific storm deposited 1.16 inches of rain on the course in 90 minutes. But Derby Day itself was on 6 June that year, and the afternoon's racing was held in pleasant weather, although the going was still on the soft side of good. The following day, however, another deluge of rain and hail struck the course and the Coronation Cup had to be cancelled.

John Bird's chronology of Derby Day weather reveals some interesting contrasts over the years. For the first hundred years or so the race was held some two to three weeks earlier than it is now – around the middle of May – and there were some rather wintry ones during Victorian times. On 15 May 1839, for instance, a biting east wind brought squalls of snow, sleet and hail throughout the day, and the race itself was run in something of a snowstorm. Similar weather occurred on 22 May 1867, with wet snow and a penetrating wind, and this inclement day probably contributed to the decision to move the race to a later date. No snowfalls have occurred during Derby week this century, although 1975 came close. Snow showers occurred on Monday 2 June that year as far south as the Thames valley, and hail fell heavily over Epsom Downs that day, while on the early morning of Derby Day itself (4 June) the course glistened under a sharp ground frost, although the afternoon was fine and dry.

There have also been some pretty soggy Derbys over the years. On 18 May 1820 heavy early-morning rain was accompanied by such a ferocious gale that the tents and booths that had just been erected around the race course were demolished. There was one horse that the wet conditions really suited that day, and that was the appropriately named Sailor, which duly won. Ten years later, on 27 May 1830, heavy rain and hail battered the participants, and there were 13 false starts before the Derby finally got under way an hour after its scheduled time. Charles Dickens was a regular visitor, and after one particularly wet day – 20 May 1863 – he wrote: "Whilst last year it was iced champagne, claret cup, and silk overcoats, now it ought to be hot brandy and water, foot baths, and flannels." The railway station at Epsom he memorably described as "an oasis of boards in a sea of mud".

In 1891 the race was actually run in a torrential downpour such that the jockeys were all a couple of pounds overweight by the end of the race, thanks to all the water their clothes had absorbed. Three very wet years occurred in succession between 1924 and 1926; by the 1920s motor-buses and charabancs were beginning to put in regular appearances on the Downs, but the waterlogged conditions during these three years meant that many vehicles became mud-bound and had to be towed out of the quagmire to the safety of the metalled roads. Attending the day's racing in such frightful weather must have been quite an endurance, and the chaos and delays at the end of the day must have caused some very frayed tempers.

18. Weather People

The history of weather study comprises three main strands: observing and recording, pure science, and weather prediction. Each could just about exist on its own as a valid field of study – but only just – so this is very much a case of the whole of weather science being greater than the sum of its parts.

The people involved in these three strands often have sharply contrasting interests and skills, and they don't always get on terribly well together. For example, some professional researchers look down on the forecasting fraternity for being academic "failures", although many of the forecasters have developed practical skills over the years that the pure scientists could not hope to equal. There are also elements among both the academics and the forecasters who despise the amateur weather observers because they are "non-scientists" and because their simple enthusiasm might be seen to detract from the seriousness of the scientific discipline. Many of those on the non-scientific side of the divide look up to the highly qualified scientists, but some recognize that a sizeable proportion of the professional community have quite extensive blind spots when it comes to knowledge of the British climate, the history of meteorology, and an aesthetic appreciation of their subject. Indeed, many amateurs have a much deeper enjoyment of the beauty and excitement of the skies and the weather than do their professional counterparts, some of whom – if pushed – would probably admit that they weren't really interested in the weather (as distinct from meteorology) at all.

All these generalizations are dangerous . . . and unfair on the majority of meteorologists, both amateur and professional, who actually get on together so well that they have coexisted within the confines of the Royal Meteorological Society since its inception almost a century and a half ago. Indeed, there are many examples of overlap between the strands: the Assistant Director of the Meteorological Office, for instance, who kept a rain gauge in his back garden and measured the rainfall every day of his professional career, and the professor of meteorology who was so entranced by the beauty and mystery of the clouds that his enthusiasm eventually grew to become his main research field. There are, of course, a handful of people who do not really fall into any of the three categories, including, it has to be said, writers and broadcasters on matters meteorological.

Over the years, the emphasis between the three strands has changed dramatically. Up to and including the whole of the 19th century, the vast majority of those who called themselves meteorologists were observational scientists who measured and analysed, classified and catalogued, compiled tables and drew maps. Thus our knowledge of the range of weather phenomena, their averages and extremes, and the distribution of weather in time and space, grew rapidly. There were a few mavericks who wanted to put all this new information to use and therefore tried hard to predict what the weather would do in the future – rarely with any sustained success. These were the forerunners of the forecasters, but they could hardly be called pure scientists because the forecasting rules they devised were empirical. Pure science remained in the sphere of the physicist, for nobody then imagined how effectively physics and mathematics could be applied to predict the future behaviour of the atmosphere. During this period, professional meteorology was largely confined to observatories – usually linked to universities – and also, from 1854, to the embryonic Meteorological Office.

The trigger for the first real movement away from observational science was the development of aviation during the first quarter of the present century. It was a gradual change at first, but it was given enormous added impetus by the First World War, when the uses of aircraft together with the need to monitor the behaviour of poison gas meant that for the first time in its history meteorology became – for better or for worse – militarily institutionalized. This led to a burgeoning in the numbers of forecasters and to a rapid expansion into meteorological research, particularly into the behaviour of the upper atmosphere. At the same time the requirement for weather observations increased, in order to feed the appetite of the forecasting offices.

One of the very few true visionaries that weather study has ever produced was Lewis Fry Richardson, a physicist by training, who stumbled into meteorology as a result of the exigencies of the Great War. He was the first person who really understood that the key to the problem of weather prediction lay in mathematics, and he published his ideas in 1922. All you needed to do was to measure the distribution of pressure, temperature and moisture throughout the whole atmosphere, apply the equations of motion, and integrate the results over the requisite period of time. There were only two problems. How do you measure the entire atmosphere? And how do you carry out the calculations sufficiently quickly to produce a forecast before the event? These are problems that have exercised the acutest meteorological minds since, tremendously aided by the advent of the computer, and Richardson's vision has now been – in large part – realized. Incidentally, Richardson used the word "computer" in his work, but to him a computer was a sort of mathematical clerk whose sole purpose was to carry out calculations. He envisaged a gigantic building containing 64,000 such computers all carrying out different mathematical tasks, and all contributing to this numerically based forecasting exercise.

Between the wars, the first chairs of meteorology appeared in British universities, while the professionalization of all aspects of weather science became almost total as the Met Office absorbed the observatories and such independent bodies as the British Rainfall Organization. Forecasting became pre-eminent, but in spite of Richardson's work, forecasting skills were still largely empirical, and the forecasters' understanding of the physics of the atmosphere remained qualitative. Even so, the more geographical aspects of the subject retained a high profile, with a large climatology section within the Meteorological Office as well as some famous names in the universities. These were typified by Professor Gordon Manley, whose knowledge of the British climate was unsurpassed, and whose magnum opus was his Central England Temperature series, a work of immense scholarship that brought together all sorts of different temperature observations from different locations and kept to different standards of accuracy, and simplified them into a homogeneous series of average monthly temperatures extending back to 1659.

Latterly, the automation of many weather observation tasks by the advent of satellite, radar and other remote-sensing technology, has led to a rapid decline in the numbers of professional weather observers. And in the last ten years the centralization and automation of forecasting processes is having the same result on professional forecasters as well. Meteorology as a source of institutional employment is shrinking, but at the same time there is the beginning of an explosion in commercial activity, and this, surely, is the main indicator of what will happen to meteorology and its people over the next two or three decades. The mavericks of the 1970s and 1980s were those who exploited their meteorological knowledge and communication skills in the private sector, in competition with the institutions. Will they form the meteorological establishment by 2025?

ADMIRAL FITZROY

Official weather forecasting in Britain began in 1859 in direct response to a disastrous gale. The embryonic Meteorological Office – at that time it was called the Meteorological Department of the Board of Trade – had already been in existence for four years under the able leadership of the irascible Admiral Robert FitzRoy. FitzRoy had earlier captained HMS *Beagle*, on which Darwin sailed on his famous expedition to South America and the South Pacific, and had subsequently been governor of New Zealand, before devoting himself to meteorology. In those very early days the job of the Meteorological Department was merely to encourage regular weather observations on land and at sea, and to collect and analyse the results. But FitzRoy soon began to feel he was little more than a glorified clerk filling in ledgers, so between 1857 and 1859 he tried to put the increasing mass of observational material to use by constructing daily weather maps.

On the night of 25/26 October 1859 the sailing ship *Royal Charter* had finally reached home waters on the last stage of its two-month-long journey from Australia to Liverpool. She was carrying half a million pounds' worth of gold bullion as well as over 400 passengers. At the same time, a deepening depression centred over the Bristol Channel around midnight was travelling steadily northeastwards, eventually reaching the North Sea later on the 26th. The strongest winds, according to FitzRoy's analysis, blew from due north down the length of western Britain, and particularly through the North Channel, the Irish Sea and St George's Channel, and may have averaged approximately 60 mph, with gusts perhaps as high as 100 mph. On the other side of the depression a southerly gale caused extensive damage to Brighton pier. The *Royal Charter* had reached the north coast of Anglesey late on the 25th when the captain decided the wind was too dangerous to proceed any further – during the final leg from Anglesey to Liverpool the wind would have tended to drive the ship onto the coast. He therefore dropped anchor, intending to ride out the storm. The details of what precisely happened are not known, but the upshot was that the *Royal Charter* foundered at around 3 o'clock in the morning on the 26th, just off the small fishing village of Moelfre, with the loss of almost all passengers, crew, and cargo. By first light the ship had sunk, and local people had managed to bring ashore just a handful of survivors. Other vessels at anchor further west survived the gale with no apparent difficulty. It was not a record-breaking storm in any meteorological sense, but its repercussions in the meteorological world were unparalleled.

In response to the storm, FitzRoy compiled a report for his political and scientific masters, making extensive use of the network of weather observations that he had been developing, and incorporating several of his newfangled weather maps. This enabled him to illustrate – sufficiently clearly for even the simplest of political minds – how the storm developed over the preceding few days, and how it was clear, 12 to 24 hours before the event, that a severe gale was about to sweep British coastal waters. Weather historians now believe he massaged some of the data in order to emphasize his case.

As a direct result, the British Association (who acted as scientific supervisors to the Meteorological Department on behalf of the Board of Trade) recommended to the President of the Board of Trade that FitzRoy's office be instructed to make use of the new electric telegraph to warn British coastal areas once gales were in existence. This service gained official authorization in June 1860, and was instituted in early September. If a southerly gale were blowing at, for instance, Plymouth, then a "south cone" would be hoisted at harbours and coastguard stations up and down the land to let seafarers know that gales were in the offing. But FitzRoy was an experienced mariner, and he knew that weatherwise sailors could spot the signs of a distant storm long before it arrived, so with his scientific background as well he felt he could do the same sort of forecasting for land areas.

Moreover, in the USA and France newspapers were already carrying "indications" of the weather to come.

The Admiral was not to be left behind, and by the following February he was issuing to shipping forecasts (or "cautionary notices") of gales not yet in existence. Within months his efforts had extended to the forecasting of ordinary weather, undoubtedly outside the brief given him by the Board of Trade, and from early in 1862 these forecasts began appearing in *The Times*. But his efforts were not very good, and periodically brought forth public ridicule. By March 1864, questions were being asked in the House of Commons, and the following winter the Editor of *The Times* finally lost faith and decided to discontinue the service. Poor FitzRoy went into a deep depression, and with personal problems, failing health and increasing deafness adding to his misery, he took his own life at the age of 60 on 30 April 1865 by cutting his throat with a razor.

JAMES GLAISHER

James Glaisher was a near-contemporary of FitzRoy, and one of his severest critics. He was a robust man who lived to a great age – he was almost 94 when he died in 1902 – and probably contributed more to the development of meteorological science and the organization of the meteorological community than anyone else in the 19th century.

He was always bound to be some sort of scientist, and, having been born in Rotherhithe, just a stone's throw from the Royal Observatory at Greenwich, he used as a small boy to pester the Observatory's staff to explain their work. His first job was as a surveyor in Ireland, where his many hundreds of hours of outdoor work gradually brought to his notice the endless variation of the sky, the patterns of clouds, and the shapes of snowflakes. This growing interest in meteorology remained just a hobby for many years, and although he returned to Greenwich in 1835, he was involved primarily in the study of astronomy and magnetism for a further five years. Upon the reorganization of the Observatory in 1840, however, he was appointed Superintendent of the Magnetical and Meteorological Department, where he remained for 34 years until his retirement.

Glaisher's work touched on a great many aspects of meteorology over the years, but there was one area that made him famous, and another for which he ought to be remembered by present-day meteorologists. He came to the attention of the newspapers, and therefore the general public, when he was persuaded by the British Association to become actively involved in a programme of balloon ascents to study the upper atmosphere. The purpose was to learn more about the distribution of temperature, pressure and moisture at various levels above the ground, and especially immediately above and below layers of cloud. These airborne studies took place between 1862 and 1866, and a further series of low-level observations was made in a tethered balloon from Chelsea in 1869. The most notorious of these ascents was one of the earliest, having been made on 5 September 1862 from Wolverhampton. The balloon reached a height of over seven miles (37,000 feet) where the temperature must have been close to −40 or −50 °C (−40 to −58 °F). Glaisher himself lost consciousness at a height of 29,000 feet, and his companion, Mr H. Coxwell, who was there to fly the balloon, was unable to operate the release valve with his hands because they were totally numb with the cold. Eventually he managed to seize the cord with his teeth and had to bob his head as hard as he could several times before the valve finally opened and the balloon began to descend again.

The other aspect of Glaisher's work was rather less exciting, but he was the key figure in establishing a network of weather-observing stations, and just as important, in fixing strict standards of instrumentation and observational practice. It is instructive to read his own words on how this all came about.

> The Registrar-General had published the mean temperature at York as being five degrees higher than at London; I wrote to him telling him of the physical impossibility of such being the case, and he then told me that he had no one in his office who could reduce the observations, and no one who could prove whether they were correct or no. I had already become desirous of ascertaining the general accuracy of meteorological instruments in general use. For that purpose I had gone all over England, Ireland and Scotland to see the best observers who were taking observations, and I found that of the thermometers the most accurate were three-tenths of a degree wrong at 32 °F, and three degrees wrong at 90 °F, and the barometers were very frequently a quarter of an inch in error. Mr Sheepshanks, in 1840 and 1841, brought out his standard thermometer, and I endeavoured to bring into general use instruments very nearly free from errors. The instrument makers worked with me, and the consequence was that at the time the Registrar-General spoke to me, I knew a large number of persons who would take observations; and, knowing a good number of Cambridge men, I thought that clergymen would unite with me and would help in establishing a system of truthful observations. Thereupon I travelled over the country and induced some fifty or sixty gentlemen of education and position to engage in the toilsome work of daily observations, and they have done so for these thirty years. I feel a great pleasure and pride to think that I have successfully organized a system which had always previously failed.

He received no remuneration for his efforts at first, and, although he was granted a small annual sum from 1854 until 1875, he thereafter continued to work for the last 27 years of his life with no payment at all. Throughout this period he took a great interest in the supply of accurate instruments to the observers, and he continuously badgered the makers of the thermometers, barometers and rain gauges to exercise the greatest possible care in the manufacturing process. Most were sent to him for examination and verification before being sent on to their ultimate destination, and for many years Glaisher was the sole recognized authority for instrument verification.

During the time when he was developing his network, he gradually came to see the benefits of establishing a nationwide meteorological society, where observers and others interested in the subject could exchange ideas, talk about their pet theories, and generally expand their knowledge. Thus on 3 April 1850, along with nine others of like mind, he set up the British (later the Royal) Meteorological Society, which continues to thrive to this very day. For 21 of the first 23 years he was the Society's Secretary, and without his unstinting efforts it would probably have folded several times during its first two decades.

Not all of James Glaisher's ideas were altruistic. He had certain commercial interests outside meteorology that allowed him to carry on his scientific work without the fear of financial embarrassment. Glaisher felt that FitzRoy was trespassing on his territory when the latter established his network of coastal observation stations, and in response to FitzRoy's pioneering weather forecasts, Glaisher immediately established "The Daily Weather-Map Company", whose object was to mass-produce and sell daily weather charts to the general public. It failed miserably to get off the ground, and only two weather maps were ever published.

GEORGE JAMES SYMONS AND BRITISH RAINFALL

George James Symons, who lived from 1838 to 1900, was the third of the three big names who dominated meteorology in England during the 19th century. He personified the Victorian approach to scientific study, being at once strongly individualistic, single-minded to the point of stubbornness, and exceptionally diligent.

Above all else he will be remembered for single-handedly

establishing a huge network of rainfall-recording stations across the British Isles, and publishing the results of the observations in an annual volume entitled, unsurprisingly, *British Rainfall*. This was a full-time occupation for the last 37 years of his life, and for many years he was assisted by his wife. His efforts were paid for by subscription sale of his publications, and by substantial donations – which he was never afraid of encouraging. For instance, in the 1888 volume he wrote under the heading "Finance":

> The subscription list . . . shows that the few continue good friends and true till death (no one but Miss Nunes ever arranged to help us afterwards) – but it is still the few. There are others contributing smaller sums, but there are over a thousand who contribute nothing. Two of these, who each professed to be deeply interested in rainfall but who had never contributed sixpence, died last year; irrespective of landed estates the personalty [personal property] of one was sworn at nearly a quarter of a million; the other, I think, was more.

Symons had a sound early education, and quickly showed an interest in the natural world, beginning his own observations of the weather and building a mercury barometer while still a child. He was a rather precocious youth, joining the British Meteorological Society at the age of 17, and presenting his first scientific paper to the Society a year later. He soon met both Glaisher and FitzRoy, becoming one of Glaisher's weather reporters in 1857, and working in FitzRoy's Meteorological Deparment between 1860 and 1863. But notwithstanding his full-time occupation, he still found time and enthusiasm to carry out his own researches into rainfall, and in 1861 he published his first collection of rainfall statistics, covering the previous year's rain at 168 places scattered across the country. The following year he received records from almost 500 contributors, and in the year of his death his British Rainfall Organization (the BRO) comprised 3500 rainfall stations. In 1919 the BRO was subsumed into the Meteorological Office, which continues to coordinate rainfall observations in the UK. Alongside the annual volumes of *British*

Rainfall, Symons also published a monthly magazine, entitled with similar imaginativeness *Symons's Meteorological Magazine*, and this acted as a sort of family newsletter amongst his rainfall observers, helping to keep the network together. The *Meteorological Magazine* was also taken over by the Met Office in the early 1920s, and continued as a monthly journal until 1993.

Many other countries attempted to emulate Symons's work, but none succeeded to the same extent.

It is thanks to the interest in rainfall observation that Symons encouraged in all parts of the British Isles that we take for granted that the mountainous western and northern parts of our islands are wetter than the plains of central and eastern England. But 17th-century scientists were much perplexed by the wetness of our mountainous regions, especially in the winter. They knew that cold air holds less moisture than warm air, and they knew that mountains are colder than plains, so they deduced that mountainous areas should be relatively dry, particularly during the winter when the temperature is lowest.

Observations to determine just how rainy our highlands are began even before Symons' time. One of our earliest rainfall records was maintained at Kendal on the southern fringes of the Lake District, but in 1836 a gauge was set up at Esthwaite Water above Windermere, and over the next few years a certain John Fletcher Miller, a Whitehaven doctor, established other rain gauges at Ennerdale Water, Wasdale Head, and ultimately in the hamlet of Seathwaite, near the head of Borrowdale.

His first year's catch at Seathwaite of over 150 inches was more than twice the Kendal figure, and it so astonished him that a second gauge was introduced, to act as a check. Miller's rainfall observer was a fellow called John Dixon, who actually lived in the hamlet, and from his knowledge of local weather conditions, he advised the good doctor to place rain gauges on the fells above Borrowdale. From these and subsequent experiments we have learnt a great deal about the wettest part of England. Long-term averages for 1941–70 give Seathwaite 123 inches per year,

Stockley Bridge 140 inches, Styhead Tarn 159 inches, Sprinkling Tarn 169 inches, and Sty Head 170 inches.

In the mid-1980s, records from these historic stations were officially terminated, in the name of "rationalization". Such is progress.

ALEXANDER BUCHAN AND THE BEN NEVIS OBSERVATORY

If FitzRoy, Glaisher and Symons were the pre-eminent personalities in English meteorology during Victoria's reign, Alexander Buchan stood head and shoulders above all others in Scotland. It is a shame that he has been remembered during the 20th century for Buchan's Warm and Cold Periods, the result of an investigation into Edinburgh temperatures over a period of several decades. He certainly never intended the extremely cautious results of his study to become public property.

Buchan's paper, presented before the Scottish Meteorological Society in 1867, noted "a tendency" towards the occurrence of cold and warm spells at certain specified times of the year in Edinburgh; he never claimed any regularity, nor any validity outside Scotland's capital city. Buchan's "periods" achieved notoriety quite fortuitously. In 1928 a Bill was debated in Parliament, the intention of which was to fix Easter to the second weekend in April. Some MP – unknown to the meteorological journals but probably traceable through Hansard – objected on the grounds that this clashed with Buchan's Second Cold Period. The reference aroused interest in the Press, and when, coincidentally, every one of the Buchan Periods arrived more or less on schedule in 1929, the interest grew into such enthusiasm that one newspaper suggested that Buchan be canonized and made the patron saint of the British weather. Heaven knows what the tabloids would make of such a story today.

Needless to say, the majority of the spells failed to appear at their appointed time in subsequent years, and thus interest in them waned, without quite disappearing altogether.

A separate Scottish Meteorological Society existed from 1853 until 1923 when, following a decline in activity and financial problems, it merged with the Royal Meteorological Society. For 47 of its 65-year life, from 1860 until his death in 1907, Alexander Buchan was the Society's Secretary and its driving force. The most remarkable achievement of the Society was the establishment of an observatory on the summit of Ben Nevis, in order to advance the knowledge of atmospheric conditions at high levels. Funds from government sources and from public donations were sought and acquired, and the observatory buildings, together with a bridle path, were constructed during 1883. While the path was being cut, masons were cutting out granite blocks from the summit of the mountain itself in order to build the observatory. It was manned continuously for 21 years, with hourly readings being made of temperature, pressure, humidity and wind.

Before the observatory was built, the Scottish Meteorological Society enjoyed the services of Mr Clement Wragge, an extraordinary eccentric who climbed the Ben daily in 1881 between the beginning of June and the middle of October, carrying out all sorts of weather observations en route, as well as at the summit, while Mrs Wragge, resident in Fort William, carried out simultaneous observations at sea level. Wragge's daily round trip took eleven hours, and must have been thoroughly exhausting even for the fittest of men when the weather was rough, as it so often is on Ben Nevis. His efforts continued during 1882 and 1883, and the Society awarded him a gold medal for "his great skill in organizing the work, his fertility of resource in emergencies, his indefatigable energy and his undaunted devotion to his work". During 1883 pigeons were used to carry the details of the weather at the summit to Fort William, whence the information was telegraphed to London.

Towards the end of the century, the observatory accounts began to show increasing deficits each year in spite of continuing public support. But with government resolutely refusing to provide more than a pittance of a grant – £350 per year as against annual running costs of around £1000 – the Ben Nevis observatory finally closed down on 1 October 1904. Even to this day the ruins of the observatory buildings are still visible just below the summit, and much of the bridle path is still in reasonably good condition and remains the main footpath up the Ben. The observatory itself is commemorated in the names of features in the great northern cliffs of Ben Nevis – Observatory Gully, Observatory Buttress, Observatory Ridge, and Gardyloo Gully, down which the residents of the observatory tipped their rubbish . . .

JAMES STAGG AND THE D-DAY WEATHER

If ever the expression "dour Scot" was invented with someone especially in mind, it was for a gentleman called James Stagg. Born at Dalkeith, just outside Edinburgh, in 1900, he became a prominent meteorologist during the 1920s and 1930s, leading a polar expedition to the Canadian Arctic in 1932 and working subsequently in Iraq for several years. In the late 1930s he was promoted to the post of Superintendent of Kew Observatory, then a senior research position in the Meteorological Office, which is where he found himself at the outbreak of the Second World War.

His name has become well known outside the meteorological community because in November 1943 he was appointed chief meteorological adviser to General Eisenhower. His task was to liaise between, on the one hand, the multifarious groups of weather forecasters who were involved in the planning for Operation Overlord, and on the other hand, Supreme Headquarters Allied Expeditionary Force, and to provide clear guidance to the decision-makers. There was a good deal of resentment amongst US military meteorologists at the appointment of a Brit to be adviser to an American general, but it was surely logical that someone with decades of experience of weather forecasting in the European arena should have got the job.

The anger simmered on as 1944 arrived, the focus of resentment shifting from Stagg's nationality to the fact that he was also a civilian. This was put right when he was very publicly given the rank of Group Captain during that April.

The problems faced by forecasters during the war were enormous. In the 1930s, weather forecasting had been as much an art as a science. Not only were computers and satellites undreamt of, but comparatively little was known of what was happening in the upper atmosphere on a day-to-day basis. Predictions were made almost exclusively (at least in European meteorological services) from sea-level weather charts with their isobars and fronts, and these were drawn up using weather observations from land stations and ships scattered across the planet. After 1940, of course, there were practically none of these weather reports over the European continent available to Allied forecasters, while ship reports became less frequent. The Met Office, quite sensibly, did not attempt to make detailed forecasts beyond 24 hours, although they usually gave a very broad-brush outlook for a further day ahead. But Eisenhower insisted on five-day outlooks. These were, naturally, little more than guesswork, and Stagg was uncomfortable passing on these guesses.

Another of his problems was how to handle the American forecasters. Meteorology was thriving in American universities by 1940, but "official" forecasting over there was stuck in something of a backwater before the war. This is not simply a British view. A well-known American meteorologist, reviewing the history of US forecasting for the *Bulletin of the American Metorological Society* in the early 1980s, described "the sad state" of forecasting between the wars.

June 1944 was one of the worst Junes of the century as far as the weather was concerned. The Americans had wanted to launch the Normandy invasion as early as possible, and certainly no later than May, while the British preferred August. In the event, May

1944 was a quiet month with long spells of settled weather and light winds, but conditions changed dramatically at the beginning of June. The decision to delay D-Day from 5 to 6 June was not terribly difficult, because the weather began to deteriorate sharply during the 4th, and the forecasters were confident in their prediction of a very rough day in the Channel. There was, however, much disagreement about the forecast for 6 June. American meteorologists said the weather would improve rapidly to provide at least three days of good weather, while their British counterparts said that any improvement would be very short-lived. The "weather window" was, in fact, not even as good as the British expected, and led to serious problems during the landings on two of the American beaches, but mercifully it was just good enough.

THE ROYAL METEOROLOGICAL SOCIETY

Many of the best-known names in meteorology in the 19th and 20th centuries have involved themselves in their spare time in the activities of the Royal Meteorological Society. Its activities are run by an elected council from which the officers and the president are drawn. The president normally serves a two-year term.

Today, the Society has a membership of approximately 3,000; a substantial minority of the members work at the Met Office, but the universities provide the largest number, including lecturers, researchers, and students. There is also a growing number of members working in private-sector meteorology and oceanography, and the Society is proud that it continues to have a large and active non-professional membership. The activities of the Society include specialist meetings and lectures, and there are all-day Saturday meetings aimed at the broad membership and especially at weather enthusiasts. Field-study courses are also offered, and much effort is directed towards fostering weather study in schools – both at secondary and primary level.

One measure of a society's activity is the quantity of material it publishes, and on this basis the Royal Meteorological Society does rather well. It produces four journals covering an enormous range of interests in the subject, and also a quarterly newsletter. The journals include the popular magazine *Weather*, the *International Journal of Climatology*, the *Quarterly Journal*, which is one of the leading publications for meteorological research in the world, and a new journal called *Meteorological Applications*, which is aimed at the professional meteorologist, especially practising forecasters.

Several prizes are awarded each year for various categories of outstanding work in meteorology, not just for high-powered academic achievement, but also for popularizing meteorology in the wider world.

The British Meteorological Society was founded in 1850, although two short-lived forerunners had been active in the London area from 1823 onwards. The new national Society arranged meetings and published research work in journal form, these *Proceedings of the Society* eventually giving way to the *Quarterly Journal* in 1871. In 1866 Queen Victoria signed the Charter of Incorporation, although the name-change to Royal Meteorological Society did not happen until 1883. The first president was S.C. Whitbread (1797–1872) of the well-known brewing family – indeed he was a director of the firm. But the driving force behind the Society's activities was James Glaisher, who was honorary secretary from 1850 until 1872. As noted above, a separate Scottish society existed from 1853 onwards, but in 1923 the two societies merged. *Weather* magazine, with a circulation of some 5000, was first published in 1946, and in recent years there have been a number of special issues, notably on the Great October Storm of 1987. An annual series of popular lectures began in 1949, the field-study courses started the next year, followed in 1954 by the first two-day summer meeting. All these continue to this day. The Society's most recent publications, the *International Journal of Climatology* and *Meteorological Applications*, began in 1981 and 1994 respectively.

The Society is keen to cater for the interests of its amateur contingent, and hopes that the numbers of weather enthusiasts who are members will continue to grow during coming years as leisure time increases. Meteorology, as much as any of the sciences, is intimately involved with the natural world, and amateur meteorologists are sometimes more aware of the aesthetic aspects of the subject than their professional counterparts. Weather observation, statistical analysis, and research into the history of weather and its study, are all areas where the amateur really does come into his or her own.

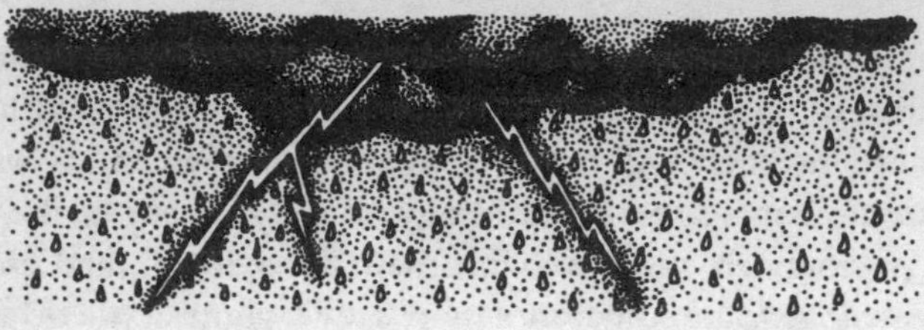

19. Worldwide Weather Disasters

For a small country we certainly have our fair share of weather disasters – rather surprisingly when you consider that our climate is so temperate and broadly speaking not given to excess. We certainly have quite a windy climate, perched as we are on the periphery of the stormy North Atlantic. But perhaps it has rather more to do with our long coastline, and the heavy concentrations of population and industry around that coastline and alongside flood-prone rivers – and possibly also because we take for granted that our weather is usually moderate and we are therefore caught out when something unexpectedly severe occurs.

It is difficult to make comparisons between different events, especially when trying to assess incomplete or inaccurate historical accounts. But in terms of the loss of human life Britain's worst two weather disasters were the Great Storm of November 1703, when some 8000 perished, more than half of them at sea, and the Great London Smog of December 1952, when it is estimated that the appalling levels of pollution contributed to an increase in the mortality rate in London and the Home Counties during the first half of the month by around 7000, although it has to be said that the majority of these should really be called indirect casualties as they were already chronically ill – many of them terminally so. One should also mention the repeated gales of late summer and autumn 1588, when the Spanish Armada was first scattered and then sunk, resulting in the loss of at least 4000 (predominantly Spanish) lives. The North Sea storm surge of 31 January and 1 February 1953 caused the most widespread flooding in recent times and 307 peopled died in eastern England, most of them drowned, although the death toll was much higher in the Netherlands. Several similar North Sea floods had occurred before, with those of December 1717, November 1530, and January 1362 outstanding, although no reliable record of casualties exists. The Aberfan disaster of October 1966, when a spoil heap collapsed and a huge wave of slurry enveloped the village school, was probably triggered by weeks of heavy rain across the whole of the Welsh valleys area, although the prime responsibility lay with human beings. There were 144 fatalities, most of them children. Smaller disasters are no less horrific to those directly affected, or to anyone with friends or relatives who have suffered, but we may mention amongst these the Lynmouth Flood of August 1952 (34 dead), the Fastnet Race disaster of August 1979 (20 dead, including 15 participants in the race who drowned), the Great Storm of October 1987 (19 deaths), and the Burns' Day Storm of January 1990 (47 dead).

Here in northwest Europe we do not experience the extraordinary violence of a mature tropical hurricane, or the catastrophic destructive power of a fully grown tornado. Nor do we have to endure the extremes of cold, heat, and drought that are part of the fabric of the global climate, and which other nations may regard as "normal", notwithstanding the periodic loss of life that may occur. But we have already noted that the storm surge of 1953 was much more destructive on the continental side of the North Sea, and a glance through the historical records shows that this is almost invariably the case. Indeed, some of the biggest weather-related death tolls through history have occurred along that vulnerable stretch of lowlands extending from Denmark and northwest Germany, through the Netherlands and Belgium, to the Dunkirk/Calais district of northern France. The climate of the early Middle Ages appears to have been particularly vigorous, with the frequency of gales higher than at present, and the limited records from those times provide tantalizing glimpses of the extent of these human tragedies – not just the huge loss of life, but also the permanent loss of vast tracts of land to the sea. The earliest useful estimates of casualties concern the floods of January 1281 in the Netherlands (between 40,000 and 80,000), December 1287 in Germany (between 25,000 and 50,000), in January 1362 (between 25,000 and 100,000), and in November 1421 (again between 25,000 and 100,000). The 1362 event was the one known as the *Grote Mandrenke* or the Great Drowning, described in an earlier chapter. The All Saints Flood of 1/2 November 1570 was probably the worst of the lot in terms of lives lost, as the heavily populated lowlands of the Netherlands were worst hit, and the cities of Amsterdam, Rotterdam and Dordrecht were totally inundated. Contemporary chroniclers estimated that between 100,000 and 400,000 were drowned. Such phenomenal disasters clearly concentrated the minds of the authorities in those at-risk countries, the system of sea defences and drainage channels was improved and improved again, dykes were raised and strengthened, and loss of life after roughly 1790 was much reduced. Meteorologically speaking, the gale and storm surge of January 1825 were even more ferocious than those of 1570 and 1362, there was widespread gale damage and considerable flooding, although the death toll was under a thousand. North Sea storms have not stopped happening, and this catalogue of death and destruction just goes to show how much modern western European society takes for granted the sea defences around North Sea coasts, and also how valuable our present-day storm-warning systems could be.

When we look further afield, famine succeeding years of drought arguably accounts for the greatest numbers of weather-related deaths, although unquantifiable social and political factors are probably just as important as the drought itself. Some estimates suggest that between 1 and 2 million died in the Indian famine of the mid-1960s, and a similar number succumbed in

Ethiopia, southern Sudan, and neighbouring countries during the 1980s.

China suffers from natural disasters more than most, particularly from earthquakes and floods, and rather like northwest Europe the death rate is so high partly because of the population density and partly because of the extent of the disasters. The huge rivers Hwang Ho and Yangtze Kiang (now called the Huang He and the Chang Jiang in modern atlases) are very prone to floods, and it is believed some 5 million Chinese have perished at the hands of these two giants. The extensive floods of July and August 1959 alone are believed to have claimed 2 million lives.

Another heavily peopled region, Bangladesh and adjacent parts of India, including metropolitan Calcutta, is also especially at risk from floods, particularly those associated with cyclones (what hurricanes are called in the Indian Ocean area). The great Bengal cyclone of October 1737 may have killed over a quarter of a million, while the devastating cyclone of November 1970 resulted in a further 300,000 drowned in Bangladesh. This was a truly devastating blow to the country, which was at the time attempting to secede from Pakistan and to establish itself as a viable independent nation.

In North America and the Caribbean, the biggest weather-related disasters usually involve hurricanes, notably the West Indian hurricane of October 1780, which killed 25,000 to 30,000, and the Galveston hurricane of 1900, which inundated that part of the Texas coast, with 6000 drowned in the town itself and up to 1000 outside. But ranked second in the USA was the Johnstown Flood of May 1899, when a flash flood raged through the small Pennsylvania township, killing 2200. Meanwhile, outside the areas affected by North Sea storm surges, the biggest European casualty list resulted from the Vienna Flood of 1342, when the Danube broke its banks and flooded large parts of the city as well as the surrounding countryside, drowning some 6000 people.

So, all things considered, maybe our little island doesn't suffer so badly after all.

HURRICANES

To a meteorologist, a hurricane is a tropical revolving storm, spawned by the warm waters of the tropics. Sometimes meteorologists forget that words have much looser meanings outside their own specialism, and that is why they looked so hurt when people laughed at them for saying that the Great Storm of October 1987 was not a hurricane. In the business, a tropical storm is given hurricane status when sustained winds in its circulation reach or exceed 74 mph, although they are "named" at a lower threshold. Strictly speaking, they are called hurricanes only in the western Atlantic and eastern Pacific; "typhoon" is the preferred name in the northwestern Pacific, which is where this kind of storm occurs most frequently, while they are called "cyclones" in the Indian Ocean and the southwestern Pacific.

After several quiet years, the late 1980s and early 1990s brought a spate of destructive hurricanes to the Atlantic/Caribbean sector. The first of these was a real "biggie" which entered the public imagination partly because of its rather harmless-sounding name – this was Hurricane Gilbert, and it happened in September 1988.

Gilbert was first spotted as a very junior tropical depression on Thursday 8 September 1988 out in the Atlantic Ocean. It swept past Barbados and across the Windward Islands during the Friday, with gale-force winds gusting to about 50 mph, with a limited amount of gale damage. It was not until Saturday morning (the 10th) that it had grown sufficiently to be classified as a proper tropical storm with strongest gusts of 60 mph and average winds of 45 mph, and it was at this point that it was christened Gilbert. At the very same time, Gilbert's predecessor, the long-forgotten Hurricane Florence, was pounding the Gulf coast of the USA around New Orleans.

On Saturday night and Sunday Gilbert gathered strength from the warm waters of the Caribbean Sea, rattling Puerto Rico, the Dominican Republic and Haiti with gusts of 120 mph – and with average winds now over 73 mph it gained full hurricane status. Jamaica had been on official Hurricane Watch from the Saturday, and sure enough, the storm hit on Monday the 12th, with 130- to 140-mph gusts howling across the island, leaving a trail of devastation. The eye of the storm travelled the length of the island, so that it felt the full force of the winds, first blowing from the north and then, after the eye had passed, from the south. And anything that withstood the wind was flooded out by torrential downpours amounting to 8 to 10 inches of rain. The death toll on Jamaica was 45; one dreads to think how high it might have been had the hurricane arrived unannounced.

Tuesday saw the full fury of Hurricane Gilbert unleashed on the tiny British Dependency of the Cayman Islands, just to the south of Cuba and to the west of Jamaica, and American meteorologists estimated highest gusts to be between 140 and 160 mph, which is almost as violent as any previously known hurricane in the Caribbean arena. The American forecasters classify fully fledged hurricanes using the five-point Saffir/Simpson Damage Potential Scale. Few hurricanes acquire Category Five status, which indicates sustained winds over 155 mph and barometric pressure in the centre of the storm below 920 millibars. The disaster potential of such a creature is laconically described as "catastrophic". Hurricane Gilbert was a Category Five storm for about 31 hours as it tracked from Grand Cayman to the north coast of the Yucatán Peninsula of Mexico. Maximum intensity was reached on the afternoon of Monday the 13th, with a central pressure of 885 millibars (almost exactly 26 inches of mercury), sustained winds of 170 to 175 mph, and peak gusts in excess of 200 mph. The Mexican island of Cozumel together with the adjacent Yucatán coastline were laid waste.

With the supply of warmth and moisture cut off during its passage across the Yucatán, Gilbert lost power by the time its eye reached the Gulf of Mexico on the other side of the peninsula. The weakening now ceased, but at least it did not now resume intensification as was feared (and as had happened on previous occasions, notably with Hurricane Allen eight years earlier). Nor did the predicted change in course take place either, so that the advised evacuation of the Texan cities of Galveston, Corpus Christi and Brownsville in the event proved unnecessary. Meanwhile, the world's media descended en masse on the small Texan town of Brownsville, which lies on the Mexican border. It was, nevertheless, uncomfortably close – less than 100 miles from the eye of the storm as Gilbert finally made landfall on the Mexican coastline – and Brownsville Airport recorded peak gusts of 60 mph, and 5.4 inches of rain.

Landfall occurred late afternoon on Friday 16 September on a sparsely inhabited stretch of Mexican coastline midway between the town of Tampico and the US border; sustained winds were still around 120 mph. Gilbert now weakened rapidly, losing hurricane status about 10 hours later, and it was finally downgraded to a tropical depression around midday on the 17th, close to the city of Saltillo. The centre of the depression passed just to the south of Monterrey with average winds of 50 to 55 mph, but it was the eight inches of rain that caused the Santa Catarina River to break its banks, the resulting flash flood claiming 200 lives. With over 3 million people living in the mountainous Saltillo–Monterrey district, the potential for far greater loss of life was mercifully not fulfilled.

Subsequently, the remnants of the hurricane drifted northwards across Texas, and were eventually absorbed by a temperate-latitude depression, which six days after Gilbert's demise brought heavy rain and high winds to the British Isles.

The nicest story to come out of Gilbert's brief but destructive life concerned a young woman who was banging noisily and angrily on the door of her local Registry Office the day after the storm. The door remained resolutely closed, the office was empty, and the registrar was not at work. When she was asked what the problem was, she revealed that she wanted to change the name of her young son who had been registered the previous week. She had called him Gilbert. . .

MORE ON HURRICANES

American weather forecasters know that summer is on the way when the list of the new season's hurricane names arrives on their desk from the US National Weather Service. Each year's hurricane season in the Caribbean/Atlantic sector begins officially on 1 June, but the final authoritative list of names is usually circulated during April.

Most of us are aware that tropical storms and hurricanes are given names beginning with successive letters of the alphabet. Names are given when the tropical storm status is reached – that is, when sustained winds in the circulation of the storm reach 39 mph. Full hurricane status is achieved when sustained winds reach 74 mph or more. On average, there are eight named storms in the Caribbean/Atlantic region each season, but there were 14 of them in 1990, and occasionally there are even more.

The habit of naming these storms dates back to the late 1940s. For many years only female names were used, since the almost exclusively male meteorological community in the USA considered girls' names appropriate for such unpredictable and dangerous phenomena. In the 1970s the growing numbers of female meteorologists began to voice their objections to such a sexist practice, and from 1978 onwards girls' and boys' names alternated. Australians, whose cyclones are just hurricanes by another name, achieved equality five years earlier.

The class of '88, when Gilbert graduated, also included Alberto, Beryl, Chris, Debby, Ernesto, Florence, Helene, Isaac, Joan, and Keith. The list is repeated every six years, but any historically memorable hurricanes will be replaced so as to avoid any confusion amongst weather historians in years to come. So when the 1988 list came round again in 1994, Gilbert's name was missing, replaced by Gordon.

A year after Gilbert left its trail of desolation through the Caribbean, another violently destructive hurricane began to rampage across a different part of the region. This one was called

Hugo, and Hurricane Hugo hogged the headlines for about a week during the middle of September 1989. It would have been difficult to match record-breaking Gilbert, let alone overshadow it, but Hugo at its peak blasted away at a steady 140 mph, with its most powerful gusts measuring 170-175 mph. It was the most ferocious hurricane to strike the north Leeward Islands since 1979's Hurricane David, and probably the most violent to cross Puerto Rico since the San Felipe hurricane of 1928. When it struck the city of Charleston in South Carolina, sustained winds of 135 mph and a tidal surge of over 15 feet made it the worst along this part of the Atlantic seaboard of the USA since at least 1954, and possibly for very much longer. After landfall, Hugo died rapidly, although (rather like Gilbert) its remnants drifted northwards into temperate latitudes, and pepped up an otherwise ordinary-looking mid-latitude depression.

After an interval of three relatively quiet years, August 1992 brought Hurricane Andrew. This was certainly a fearsome storm, about as bad as Hugo but not quite in the same league as Gilbert. It was one of those rare hurricanes that have two bites of the cherry; first of all it tore a swathe across the southern extremity of the Florida peninsula, moving from east to west, and then with undiminished intensity it headed across the Gulf of Mexico to make a second landfall on the coast of Louisiana. It was by no means the most severe tropical storm to hit the USA, but it was reported to have done more damage than any other – although this probably says more about the population increase in Florida and Louisiana in recent years, and the universal turning of blind eyes to planning recommendations in respect of erecting new buildings on hurricane-prone coasts.

Andrew was first spotted in the tropical mid-Atlantic early on Monday 17 August, when it was classified merely as a "tropical depression" with sustained winds of just 35 mph. Later that day it was upgraded to "tropical storm" status with wind speeds of 40 to 50 mph. It was heading steadily westwards, but further growth seemed out of its reach for several days. Then suddenly on the morning of Saturday 24 August it intensified sharply, and as sustained winds exceeded 73 mph it was re-classified as a fully fledged hurricane. It was now only 600 miles east of the Bahamas and heading directly for them. The eye of the storm passed over the Bahamian island of Eleuthera on the Sunday evening while Andrew was at its greatest ferocity. Winds averaged 150 mph, with peak gusts of 185 mph.

In the wake of these dramatic storms there were ominous rumblings from some climatological commentators, who claimed that they were just another illustration of the way global warming is beginning to play tricks with the earth's atmosphere. It is thought that one of the main contributors to global warming is the so-called "greenhouse effect", caused by the gradual increase in quantities of carbon dioxide and other gaseous pollutants in the atmosphere. These restrict the amount of warmth that escapes into outer space, which in turn causes a very slow rise in the planet's mean temperature. The oceans are by far the largest store of heat energy on earth, and some studies show that ocean-temperatures now average over 0.5 °C (1 °F) warmer than thirty years ago. This, according to the commentators, provides a perfect background for more frequent and more intense hurricanes. The key requirement for the formation and growth of a hurricane is a substantial area of ocean with a surface temperature of 27 °C (81 °F) or more – the warmth and moisture providing a constant supply of energy needed to fuel these storms. Such temperatures are found in the Caribbean, the Gulf of Mexico and adjacent parts of the Atlantic Ocean from June through to early November. The peak occurs in September, which is also the peak of the hurricane season.

As global warming takes hold, so the theory goes, we can expect our oceans to become a few degrees warmer during the next century or so. Thus, in tropical areas, that crucial 27 °C mark will be exceeded over larger areas, for longer periods, and in the hottest areas by a greater margin, thus supplying added energy to any incipient hurricane. To understand the impact that this may have in the Atlantic-Caribbean area, we need only look across the world to the western Pacific. A much larger area of the Pacific Ocean exceeds 27 °C, and temperatures stay near that level throughout the year in the neighbourhood of Indonesia and New Guinea. Pacific typhoons (same storm, different name) can occur in any month of the year, regularly reach the intensity of a Gilbert, and there are on average 21 per year compared with 7 per year in the Atlantic–Caribbean. Destruction on the scale seen in Jamaica in September 1988 is no stranger to countries like the Philippines, Taiwan, Vietnam, and south China.

WEATHER AND SADDAM HUSSAIN

The weather was front-page news in British newspapers on the first two days of August 1990. It would have stayed on the front pages on 3 and 4 August as well, but for Saddam Hussain invading Kuwait, because those two days marked the peak of Britain's most extreme heatwave of the 20th century.

It was almost six months before the anti-Iraqi alliance was in a position to launch a counteroffensive to liberate occupied Kuwait. As we saw from the television pictures, the local weather conditions played a considerable part during the seven weeks of war, and weather forecasters also undertook a vital, unsung role.

We naturally think of deserts as places where it never rains and where the sun constantly shines. This is, of course, a gross generalization; it is not true of the majority of deserts on the earth's surface, and it is certainly not true of the Arabian peninsula during the "winter" season. After all, Iraq and Saudi Arabia are protected from icy blasts whistling down from Siberia only by the mountain ranges of Iran and Turkey, while winter depressions from the Mediterranean with their cloud and rain periodically sweep down from the northwest. In the war zone, daytime temperatures average about 20 °C (68 °F) during January and February, while night-time readings oscillate around 8 °C (46 °F), occasionally dropping close to freezing point. Early mornings can be quite humid with fog an occasional hazard, while the average cloud cover is something like 60 per cent. In other words, the sky is more often cloudy than not. Significant rain falls on an average of four days per month – roughly once a week – usually very heavily for a short period. Hail and thunder occur about once a month, and even snow has been known, although it is very rare indeed. The prevailing wind is northwesterly, and it is occasionally strong enough to lift dust and sand from the desert floor, sometimes resulting in sandstorms.

What does this mean in terms of a typical week's weather? What were the problems the forecasters had to deal with? A Mediterranean depression approaching from Israel and Jordan is typically preceded by a very warm, humid day with thickening clouds, usually high in the sky, and a light southeast wind. As the depression gets close, the clouds will grow thicker and lower, the wind will strengthen, and what met-men call "rising sand" becomes a problem. Rising sand is a sort of low-grade sandstorm that is confined to the lowest four or five feet of the atmosphere. It is of course very unpleasant to be caught in, and it plays havoc with machinery and electronic equipment. As the depression or cold front passes through, there is usually a sudden downpour of rain, perhaps accompanied by thunder and hail, maybe lasting three or four hours, and the desert sand and dirt will turn into a sea of mud for a day or two. Violent gusts of wind often occur at this stage too, as the wind switches from a southeasterly to a northwesterly. Behind the storm the weather will clear up, but a strong nor'wester known as a "shamal" can last for a couple of days afterwards. This helps to dry out the quagmire, but once the desert is dry again, the sand will blow about in the wind. When the wind blows itself out the weather will become quiet for several days, with variable amounts of high-level cloud and plenty of warm sunshine.

It was very interesting during the period of hostilities to compare the pictures we were getting on television news bulletins with what could be deduced from the weather-satellite images. Over and over again there were mismatches, for instance the pictures of Kate Adie earnestly explaining the problems of transporting military equipment across a sea of mud through torrential downpours and under a lowering sky appeared to coincide with satellite pictures showing a thousand-mile wide cloud-free swathe covering the whole of the Arabian peninsula. It appeared that the so-called "actuality" from the war zone itself was in fact two days old. This enabled the military authorities to control the news output throughout the war, whatever the radio and television correspondents in Riyadh would have liked us to believe – indeed, the flow of information to them was being so successfully manipulated that they probably knew less about what was going on than we did in our own front rooms. When I rather naïvely pointed out the two-day delay in a radio broadcast, a polite but firm telephone call from the Ministry of Defence to the news editor suggested that the weatherman should confine his comments to the weather itself, and preferably to the weather that had already happened, and not what might happen over the next few days.

There was much concern at various points during and after the Gulf War about the possible impact of the pollution from oil fires on the global climate. This largely took the form of irresponsible and sensationalist stories in the print and broadcast media which were really no more than axe-grinding and pot-stirring from environmentalist pressure groups and tabloid editors – equally meteorologically illiterate – who suddenly found themselves, rather unusually, in the same bed.

If they had sought scientific advice, they would have learnt that the atmosphere in the Middle East is what meteorologists call "stable". This means that air tends to sink most of the time, especially in the middle and upper parts of the troposphere. When the oil refinery in the Saudi Arabian town of Khafji was bombed early in the war, we saw an impressive plume of smoke on the TV pictures, but it was also clear from these pictures that the smoke was rising about 2000 feet into the atmosphere and then it just stretched out horizontally; it could climb no higher because of the descending air currents. Similarly, when the Kuwaiti oil wells were fired by the Iraqis just before they retreated, nearly all the pollution remained in the lowest 10,000 feet of the atmosphere. It was thus prevented from reaching the stratosphere, which is what would have had to happen if the smoke and soot were to be transported around the northern hemisphere. The pollution was thus concentrated in the local area, and there was never any real chance that the Indian monsoon would be affected, let alone weather patterns in other parts of the world. True, discoloured snow was seen in the Himalayas during March and April, but even this goes to show that any pollution that escaped the Gulf area remained in the lower part of the atmosphere where all the weather systems live, and was later rained out over Iran, or snowed out over the Himalayan mountains.

CALIFORNIAN HELL

In December 1988 there was one natural disaster that made the headlines in almost all parts of the world. That was the Armenian earthquake that devastated the city of Leninakan (now renamed Kumairi) and completed razed the town of Spitak. The casuality list approached 100,000. But there was one part of the world where the story was relegated to second place by a more local catastrophe.

Southern California is well-known for the parochial nature of its news bulletins. Local news happens in the Los Angeles metropolitan area, "national" news encompasses the rest of southern California and maybe one or two important events as far away as San Francisco, and the few snippets of 'international news' usually involve what the politicians are saying in Washington DC. Any other country might as well be on the moon. But on this occasion there was some justification for the *Los Angeles Times* leading with a story about the fierce brush fires sweeping through some of the outer fringes of the metropolitan area.

On 9 December major fires raged through the Sunshine Canyon, across the Santa Susana Mountains and into Limekiln Canyon, and, fanned by winds gusting over 70 mph, secondary offshoots spread into the newly developed suburban zones of Porter Ranch and Granada Hills. Fifteen homes were burnt to the ground, 25 others were badly damaged, and 8000 residents were evacuated as the fires appeared to gain strength. A firefighting force of over 1000, gathered from various stations across the northern segment of Greater Los Angeles, struggled without success to gain control. Some 50 firefighters together with half a dozen residents were injured, almost 10 square miles of brush were consumed by the flames, and total damage was estimated at 4 million dollars. The day before, firestorms in the Baldwin Park and La Verne districts had caused 10 million dollars' worth of damage, razing to the ground 22 homes and damaging several others, and resulting in several hundred evacuations.

Such fires are a chronic hazard in the Los Angeles basin and the surrounding hills because of the occasional occurrence of hot dry Santa Ana winds in this part of southern California. The Santa Ana is a northeasterly gale that originates in the Mojave Desert and funnels through mountain passes on its way to the heavily populated coastal strip. As it is squeezed through the gaps in the moutain ranges, this wind reaches phenomenal strength and it greedily sucks up all available moisture, leaving the natural vegetation tinder-dry. It takes its name from one of these canyons – the Santa Ana Valley. It is difficult for anyone who has not experienced them to imagine just how hot and desiccating these winds are. Gusts of 80 to 100 mph may be accompanied by temperatures in excess of 40 °C (104 °F) and relative humidity of 5 per cent or less.

The December 1988 fires were not brought under control until the Santa Ana had blown itself out, although mercifully damage was relatively limited. But other occasions have brought much more widespread devastation. A ferocious Santa Ana was responsible for the fierce fires that destroyed the homes of several Hollywood stars in Beverly Hills, Bel Air and Brentwood in November 1961, and in September 1970 1000 homes and other buildings were burnt to the ground in the San Diego district close to the Mexican border. Film stars' homes were again in the front line early in 1994 when fires raged on several fronts, including Malibu and Santa Barbara.

Similar "downslope" winds have brought similar disasters to the Berkeley Hills area on the east side of San Francisco Bay. In this area the winds are simply called "northers". The university city of Berkeley itself suffered its worst tragedy in September 1923 when high winds fanned the fires across the northern outskirts at a terrific rate. Within three hours almost 600 homes had been destroyed, and the fire reached one corner of the university campus when the winds finally gave out and the fires lost their strength. After this conflagration, the city authorities undertook extensive and expensive fire-prevention work in the city, and this was proved worthwhile in September 1970 – at the same time as the San Diego disaster – when a potentially devastating fire was kept under control.

20. The Changing Climate

OUR CLIMATE SIMPLY WON'T SIT STILL

The worldwide interest over the last decade in the threat of climate change implies that until now the climate has been more-or-less static. The view of the general public in the past seemed to be contradictory – on the one hand comfortable in the notion that, within reasonably narrow bounds, the climate was reliable and unchanging, and on the other hand always complaining that "We don't have summers now like when we were kids," or "It's ages since we had a real old-fashioned winter; when was the last time we had a proper snowstorm?" There was the same attitude a hundred or so years ago. Even then, people remarked on what they perceived to be changes in the climate since their childhood, while the meteorologists of the time, still barely scratching the surface of knowledge about historical aspects of climate, believed that our climate probably had been static for all practical purposes since the end of the last ice age. That was when the 30-year standard period for climate statistics and averages was devised; 30 years was needed so that one or two extreme events did not materially affect the overall statistics, otherwise such a period was quite sufficient to describe the climate of a particular place both in the past and in the foreseeable future.

We now know a little better. We know that climate changes

are the norm, and that fluctuations in the global climate as well as the climate of individual locations have happened throughout the planet's history and will continue so to do. And we also know that these fluctuations occur on different time scales, from small and apparently (though probably not) random changes that may last a deade or two, to changes spread over tens of millions of years caused by continental drift. Very important fluctuations (some people call them cycles, but this is misleading) over periods of thousands of years – probably caused by changes in the orbit and axial tilt of the earth – resulted in the glacial periods (commonly called ice ages) during the last 2 to 3 million years. The experts refer to the whole of that period as the Ice Age, while the colder periods are called "glacials" and the warmer periods "interglacials"; we are at present in the midst of an interglacial period.

Since the end of the last glacial period in Europe, roughly 10,000 years ago, several important changes spread over periods of hundreds or even thousands of years have occurred. The so-called Climatic Optimum (sometimes called the Sub-Boreal phase) lasted from approximately 4000 BC to 2500 BC when mean temperatures in the UK were roughly 2 °C above present values and the tree line in Scotland was close to or just above the 3000-foot contour (it is now below 2000 feet). This was followed by a marked deterioration into the Sub-Atlantic phase between 1000 BC and AD 500, which was rather cooler and wetter than the present day. Since then, there have been several rather smaller fluctuations. These include a warmer but often stormy period during the early Middle Ages, when, although vineyards were scattered across England as far north as Yorkshire, repeated storm surges in the North Sea caused disastrous inundation of the low-lying coastlands of Jutland, northwest Germany and the Netherlands.There was also a much colder phase between 1550 and 1750, with frequent severe winters and poor summers – a phase commonly referred to as the Little Ice Age.

On a much shorter time scale, some of the more noticeable

changes in the British climate during the last 100 years include: a marked trend towards warmer autumns (especially Octobers) between the 1920s and mid-1970s; a change from warm springs centred around the 1940s to much colder springs during the 1960s and 1970s; and a change from very mild winters between 1900 and 1939 to frequent cold winters from 1940 to 1970. Meanwhile, summers appear to have grown drier since about 1970.

So the prospect of future climate change caused by mankind's interference with the atmosphere has to be viewed in the context of all these "natural" changes that have happened in the past, and which are probably happening at the moment. We have already noted that some of these climatic variations may have been caused by the change in the distribution of continents and oceans over the planet and by changes in the tilt of the earth's axis and the shape of its orbit. But there are many other influences, such as variations in the position of the North and South Poles, the building and erosion of mountain ranges, changes in the composition of the earth's atmosphere and in particular the quantity of the so-called "greenhouse gases" such as carbon dioxide, changes in ocean currents and variations in the effectiveness of the oceans to store heat energy, and even the impact of large meteors and cometary and asteroid residues. On a much shorter time scale, the impact of volcanic eruptions is generally accepted as being very important, but especially those eruptions that inject large quantities of sulphurous material high into the stratosphere. There are also feedback mechanisms that have to be taken into account: for instance, a change for whatever reason to a colder climate will increase the coverage of ice and snow, alter the vegetation cover in other areas, cause a drop in sea level, and affect ocean currents and wind patterns – and all these changes will have their own additional impacts on the climate.

The atmosphere is a very resilient machine, but we can see from the evidence of the glacial periods that it is relatively easy to push it over the brink from one relatively stable pattern to a very different one. The equilibrium within both the glacial and the interglacial episodes may each last thousands of years, but the "flip" from one to the other may take just a few centuries, or possibly only decades. So whatever we think of the evidence about man-induced climate change, we should be aware that we probably have the capacity to induce one of these "flips" if we are not careful. Some people say that there is nothing wrong with human-induced changes in the climate, because the planet has shown that it can cope with changes in the past. The planet may be able to cope, but can we? Our very complex societies – relying on relatively small nation states with unprecedented population density and intensive agricultural practices – are intricately dependent on a stable natural environment. *Any* change in that environment will pose incredible strains on us – especially on our ability to feed ourselves. In the past, the earth's flora and fauna coped with climate change by mass migration, an option that is hardly open to mankind in the 21st century.

GLOBAL WARMING AND THE GREENHOUSE EFFECT

Temperature levels in the lowest few feet of the atmosphere – where we live – show little change when averaged over a period of years. This is the result of a balance between incoming energy from the sun, and outgoing energy radiated from the earth's surface. The atmosphere lets nearly all the incoming radiation through to ground level, but the outgoing radiation is at a different wavelength and some of it is absorbed and re-radiated by carbon dioxide, methane, water vapour, ozone, and other so-called "greenhouse gases". This trapping of heat energy is rather like the way heat builds up in a closed greenhouse, and this is where the term "greenhouse effect" comes from. So we are actually able to live comfortably on the surface of our planet because of a natural greenhouse effect that maintains the temperature at a considerably higher level than it would be if it were possible to counteract that effect. Some experts think that this natural greenhouse effect warms the earth by between 30 and 35 °C.

Human activity has been increasing the concentration of greenhouse gases – in particular carbon dioxide – since the beginning of the Industrial Revolution, about two hundred years ago. In that time, the quantity of carbon dioxide in the earth's atmosphere has increased by about a quarter, and many scientists now believe that this is sufficient to produce an increase in the average temperature at the earth's surface. Some even think that this rise in temperature is already detectable. We have also created CFCs (chlorofluorocarbons) for use in aerosol cans, refridgeration units and packaging materials, and CFCs, as well as having a destructive effect on stratospheric ozone in polar regions, are an entirely new type of greenhouse gas. Carbon-dioxide emissions have increased mainly as a result of the burning of coal, oil and natural gas; the cutting down of tropical (and other) forests and the growth of the deserts across previously vegetated land also make a contribution. But it is industrial countries that are the biggest culprits – the USA emits twice as much carbon dioxide per person as the UK and fifty times as much as Bangladesh. Without any change in industrial processes, it is estimated that we will double the pre-1800 level of greenhouse gases by the year 2025.

A substantial increase in the quantity of greenhouse gases in the atmosphere should result in a substantial increase in the overall average temperature of the whole earth/atmosphere/ocean system. But it is very difficult indeed to determine the effect on the climate of a particular region or particular place. As the temperature rises, there are several feedback effects that come into play. For instance, the warmer the atmosphere becomes, the more moisture it can hold. The increase in water vapour enhances the warming effect because water vapour is a greenhouse gas, but the likely increase in cloud cover should reflect some of the incoming solar radiation, thus producing a cooling effect. Computer programs have been devised to model the way the atmosphere works, but although these are improving all the time they are still relatively crude, and their predictions must be taken with a wheelbarrow-full of salt.

The first serious estimate of the warming effect was made by an international panel of scientists in 1990. They calculated that, unless changes occurred in the emission of greenhouse gases, the average temperature of the earth's surface would increase by about 2 °C by the year 2050, although uncertainty meant that the figure could be as little as 1 °C or as much as 3.5 °C. Since 1990, improving understanding of how the feedback effects work and new information about the part played by the oceans suggest that the warming process will probably be rather smaller than this. But as knowledge grows over coming years, these figures will doubtless change again – probably several times, and possibly in different directions.

The other major impact of global warming is a partial melting of the polar ice caps, especially the Antarctic ice, which would have two very important effects. It would pump large quantities of cold fresh water into the surface layers of the oceans, and it would also cause a global rise in sea level. Some estimates suggest that sea level might rise by between one and two feet by the year 2050, but there are again many poorly understood feedback mechanisms that should be taken into account, and to be honest there is no real consensus on this subject.

It is important to recognize that a global-warming effect does not necessarily mean that all parts of the world will become warmer, and it certainly does not mean that the warming effect will be roughly the same across the whole planet. Here in the UK a lot of simplistic twaddle has been written and spoken by journalists, broadcasters and environmental pressure groups pointing to a change in the climate of southern Britain to something akin to that which exists at the moment in southwestern France. This, of course, encourages the general public to think that global warming might be a good thing if it brings us Mediterranean summers, citrus fruit and vineyards, and reduces heating bills. It is easy to ignore the fact that it might also wash away our beaches, flood the Fens, bring hitherto unknown pests and diseases, increase levels of photochemical smog, and deposit large air-conditioning bills on our doormats.

What should we do about it? What *can* we reasonably do about global warming when our understanding of its repercussions is so limited? Mankind probably has the power to change the earth's climate, but has little or no control over what those changes might be. It is quite possible that the computer models seriously overestimate the impact of greenhouse gases on temperature levels, in which case changes in industrial processes – which are likely to be expensive, disruptive, and politically sensitive – may turn out to be largely unnecessary. On the other hand, the models may underestimate the degree of change during the next century, with possibly frightening consequences.

We can probably make a few useful observations. There is clearly a long response time to changes in human activity; so if we need to make major adjustments to the way we generate power and use fuel, we know that it will probably take decades before the climate machine responds. It is rather like trying to make an ocean liner change direction. Thus any climate changes induced by man's activities in the last half-century are still in the system, and they will show their hand in the next fifty years – and we cannot stop them. Even now there are things that we can do to reduce emissions that are probably worth doing anyway, such as insulating buildings, making transport systems more efficient, ending unnecessary forest clearances especially in the tropics, and changing to renewable energy sources (such as wind-power, tidal energy, and solar power) if and when they become competitive. Nor would a more concentrated effort into researching and developing these alternative sources of energy cost all that much in the great scheme of things. Such low-cost changes are sometimes called "no regret" policies. We should also be aware that the planet's population is doubling roughly once every 35 to 40 years, and in the business-as-usual scenario this would accelerate the emission of greenhouse gases into the atmosphere.

THE ALTERNATIVE WEATHER FORECAST FOR THE YEAR 2050

Most newspapers, especially the tabloids, thought that a quasi-Mediterranean climate in Britain might actually be a good thing. But this was only one possibility in a whole range of possible climate-change scenarios for the middle of the 21st century. There is no certainty even that the average temperature in the vicinity of the UK will actually increase during coming decades. Indeed, when it comes to estimating what might happen to our own climate over the next half-century or so, there are probably as many different theories as there are experts.

So far, the south-of-France theory has received virtually all the publicity. Let us try to balance the books and look at a rather less attractive one.

As the earth's average temperature rises, the atmosphere will become more humid. This is because the warmer the air is, the more moisture it can hold. Thus humidity levels will become even higher in the tropics than they are now, perhaps resulting in increased cloudiness, and smaller amounts of sunshine. The reduction of sunshine together with an increase in rainfall may cause a small drop in the temperature of the tropical oceans. At the other extreme, in polar regions the cloud cover would act rather like a blanket, keeping warmth in, but there is also likely to be an increase in the amount of snowfall over the ice caps of Greenland and Antarctica. The rate of ice-melt around the fringes of the ice caps would increase, but this could well be balanced by the added snowfall, so the quantity of ice tied up in polar regions may not change significantly. Similarly, snowfall may well be heavier during winter over northern Canada and Siberia, but the *geographical extent* of winter snow cover across the Eurasian and North American land masses would probably diminish.

The result of all this is that the warm ocean currents from the tropical regions would be weaker than at present, while the cold meltwater from the polar ice caps is likely to extend over the ocean surface, in the Atlantic perhaps as far south as the latitude of the British Isles.

In winter, therefore, we would have created warmer continents and cooler oceans compared with the present day. But it is the existing contrast between warm ocean and cold continent that provides the energy driving the big Atlantic depressions, which in turn control the character of typical winter weather in northwest Europe. Thus in our new situation, the depressions become smaller and less vigorous, and they travel across the Atlantic at a more southerly latitude – along the boundary between the cold meltwater outflow and the weaker warm ocean current. If the favoured depression track is towards the Bay of Biscay and thence to the Mediterranean, our winter climate would be controlled by semi-permanent high pressure near Iceland or Scandinavia, and our winters would be colder and drier and less windy than now, with frequent frosts and fogs.

In summer, the temperature contrasts that drive the weather system are between hot continent and relatively cool ocean. Under the influence of global warming, the continents become even hotter (they are too dry for the increased atmospheric moisture to create any added cloud cover), while the oceans stay cooler because of the continuing supply of Arctic meltwater. The end result of this is that the temperature contrast becomes larger than it is at present. This will bring about bigger and stronger depressions in summer, delivering predominantly cool, windy and cloudy weather to the British Isles. We had a taste of this in 1988 – Britain had a remarkably cool and unsettled July with excessive rainfall and persistent cloud cover, while the USA and much of Europe and what was then the USSR suffered record-breaking heatwaves.

This particular theory is terribly over-simplified, and it is probably full of holes, but so there are in most other theories as well. But it could happen. So here is the alternative weather forecast for the year 2050: cold dry winters with plenty of fog but not much snow, and cool wet summers with leaden skies and blustery winds. Our farms grow oats and potatoes, while apples and pears fail to ripen as often as not. And because of the rising sea level and our inability to pay for improved sea defences (because of our impoverished economy) the Fens revert to marshland, Canvey Island is permanently evacuated, and the Thames Barrier is given a second storey. Sounds exciting, doesn't it?

THE OZONE HOLE

Ozone can be a very confusing subject. On the one hand we are told that the ozone layer in the stratosphere is a good thing, protecting us from harmful ultraviolet radiation. But on the other hand we are warned that ozone in the lowest layers of the atmosphere, where we live and breathe, is a bad thing, threatening out health. The low-level ozone is caused by the action of sunlight on car exhaust fumes so it is actually nothing more than an unwanted pollutant for which modern society is responsible, whereas the stratospheric ozone layer is a natural phenomenon.

Ozone is a comparatively rare form of oxygen, containing three linked atoms instead of the usual two. This is why it's referred to in shorthand as O_3, while normal oxygen is often called O_2. About one-fifth of the atmosphere consists of oxygen, but ozone comprises only a few millionths of it, and most of that is in the upper atmosphere – in that stratospheric ozone layer between thirty and forty miles above the earth's surface. Its most important function is to filter out a large proportion of the ultraviolet radiation coming from the sun. It removes nearly all of the very damaging UV-C wavelengths and between 70 and 90 per cent of the less damaging UV-B wavelengths. Ultraviolet rays are responsible for sunburn and snow blindness, and they contribute to the ageing and wrinkling of the skin, and also to some eye damage and skin cancers. It is believed that about

10,000 people die from skin cancer each year, and that in regions with what you might call a white-skin sunshine culture (such as California and Australia) between a quarter and a half of all cancer deaths are from ultraviolet-induced skin cancer.

Stratospheric ozone is produced by the effect of sunlight on ordinary oxygen, and given time it will eventually decay back into ordinary oxygen. Ozone is also destroyed by other chemicals that are found naturally in the atmosphere, such as oxides of nitrogen, methane and water vapour, while artificially produced CFCs (chlorofluorocarbons), used in fridges, aerosol cans, expanded polystyrene, and so on, have the same effect. The ozone layer will only remain in the stratosphere to protect us from harmful radiation as long as there is a proper balance between the generation and destruction of the gas.

Measurements made in Antarctica show a decrease in stratospheric ozone in the region above the south polar zone since the 1970s, most marked during the southern-hemisphere spring. Since the early 1990s there has been growing evidence of similar but smaller ozone loss in the northern hemisphere, also during spring. These ozone losses – the so-called "ozone holes" – not only threaten our health, but they could also have a serious impact on the earth's climate, although we do not yet begin to understand how this might happen.

What are we doing to counter the threat of ozone depletion? International concern began long before the media became interested in the subject. In 1977 the United Nations brought together a panel of experts to draft a plan of action, and in 1985 a Convention for the Protection of the Ozone Layer was adopted by a large number of countries. The main result of this was a set of rules for the control of CFCs, known as the Montreal Protocol. It might surprise some journalists that the Montreal Protocol was actually conceived so many years ago, and was not a response to the media campaign of the late 1980s. CFCs enter the atmosphere by leakage from industrial equipment and installations, by release from aerosol cans, and from refrigeration units when they are dumped or broken up. Other gases also make a contribution to the destruction of ozone – nitrogen oxides and water vapour from high flying aircraft, nitrogen compounds from fertilizers, and methane from a variety of processes. But CFCs spend much longer in the stratosphere than the other gases, so they continue to pose the greatest threat to the ozone layer.

Computer models have predicted what might happen to the stratospheric ozone layer in the future. If we continue to produce CFCs at present levels, it is estimated that there will be a 7 per cent reduction in the total quantity of ozone in the next 50 years. If CFC production were to increase by just a small amount, say by 3 per cent per year, roughly one-tenth of the ozone would vanish by the year 2050. Even if we stopped production of CFCs completely by the year 2000, ozone depletion would continue for a further 5 to 10 years and ozone levels would not return to earlier levels for another 60 years or so.

We have already looked at the worrying effects that increasing ultraviolet radiation has on human beings, and there would also be a serious impact on other living things on the planet. It is thought that roughly two-thirds of food crops could suffer some effect – some not germinating properly, and others growing more slowly. Many single-cell organisms in the surface layers of the oceans would be killed by exposure to increased UV-B radiation, and these organisms are particularly important because they are at the bottom of the aquatic food chain.

So experts believe that a reduction of only a few per cent in stratospheric ozone conentrations would lead to an increase in eye damage and skin cancers amongst both human beings and livestock, reduced production of agriculture and fisheries, reduced timber production, and the marine environment could suffer dramatic changes with unknown consequences. There could well be more photochemical smogs in industrial areas, while plastics, paints and some building materials would probably deteriorate faster than at present. Thus many scientists believe that the continued production of large quantities of CFCs poses an unacceptable risk to all human society and to the environment on which it depends. Monitoring of and research into ozone depletion must continue, and gases and chemicals that are produced on a large scale should always be examined to identify any possible detrimental effect that they may have on ozone in the stratosphere.

Global climate change, whether natural or induced by human beings, may well be one of the more prominent strands of scientific interest during the 21st century.

To the politician it will probably be one of those topics that keep turning up like the proverbial bad penny, high in nuisance value but to be avoided in terms of policy changes unless some disaster intervenes; to the journalist, the broadcaster and their multimedia successor it will doubtless be a bottomless pit of news stories, features and interactive programming, and to the more imaginative a source of great fun and mischief; to the public at large it could well be just a subject of intermittent concern and continuing confusion.

But one thing is for sure; if we are still here in the year 2100, we will still be talking about our changing climate.

APPENDIX

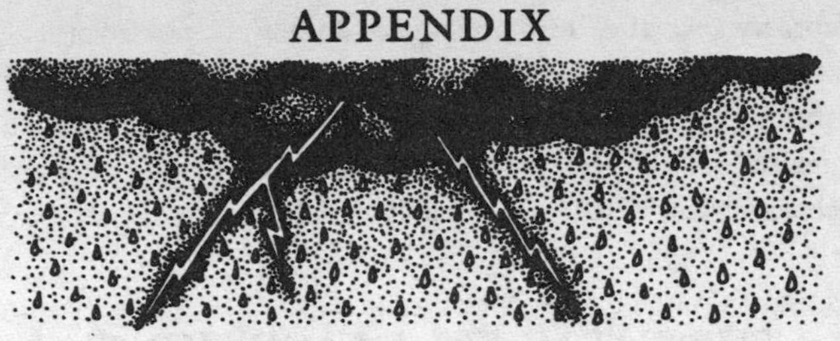

Weather Lists

THE 10 HOTTEST MONTHS SINCE 1659
(according to Gordon Manley's Central England Temperature Series)

1. July	1983	19.6 °C (67.3 °F)
2. July	1783	18.8 °C (65.8 °F)
3. July	1852	18.7 °C (65.7 °F)
4. July	1976	18.7 °C (65.7 °F)
5. August	1975	18.7 °C (65.7 °F)
6. August	1947	18.6 °C (65.5 °F)
7. July	1921	18.5 °C (65.3 °F)
8. July	1757	18.4 °C (65.1 °F)
9. July	1808	18.4 °C (65.1 °F)
10. August	1990	18.4 °C (65.1 °F)

THE 10 COLDEST MONTHS SINCE 1659

1. January	1795	−3.1 °C (26.4 °F)
2. January	1684	−3.0 °C (26.6 °F)
3. January	1814	−2.9 °C (26.8 °F)
4. January	1740	−2.8 °C (27.0 °F)
5. January	1963	−2.1 °C (28.2 °F)
6. January	1716	−2.0 °C (28.4 °F)

7. February	1947	−1.9 °C (28.6 °F)
8. February	1895	−1.8 °C (28.8 °F)
9. February	1855	−1.7 °C (28.9 °F)
10. January/	1776	−1.6 °C (29.1 °F)
February	1740	−1.6 °C (29.1 °F)

THE WARMEST OF EACH MONTH

January	1916	7.5 °C (45.5 °F)
February	1779	7.9 °C (46.2 °F)
March	1957	9.2 °C (48.6 °F)
April	1865	10.6 °C (51.1 °F)
May	1833	15.1 °C (59.2 °F)
June	1846	18.2 °C (64.8 °F)
July	1983	19.6 °C (67.3 °F)
August	1975	18.7 °C (65.7 °F)
September	1729	16.6 °C (61.9 °F)
October	1969	13.0 °C (55.4 °F)
November	1994	10.3 °C (50.5 °F)
December	1934	8.1 °C (46.6 °F)
Warmest Year	1990	10.7 °C (51.3 °F)

THE COLDEST OF EACH MONTH

January	1795	−3.1 °C (26.4 °F)
February	1947	−1.9 °C (28.6 °F)
March	1674	1.0 °C (33.8 °F)
April	1701	4.7 °C (40.5 °F)
	1837	4.7 °C (40.5 °F)
May	1698	8.5 °C (47.3 °F)
June	1675	11.5 °C (52.7 °F)
July	1816	13.4 °C (56.1 °F)
August	1912	12.9 °C (55.2 °F)

September	1674	10.5 °C (50.9 °F)
	1675	10.5 °C (50.9 °F)
	1694	10.5 °C (50.9 °F)
	1807	10.5 °C (50.9 °F)
October	1740	5.3 °C (41.5 °F)
November	1782	2.3 °C (36.1 °F)
December	1890	−0.8 °C (30.6 °F)

THE HOTTEST DAYS OF THE 20TH CENTURY

1.	3 August	1990	37.1 °C (98.8 °F) at Cheltenham
2.	9 August	1911	36.7 °C (98.0 °F) at Raunds
3.	2 August	1990	36.6 °C (97.9 °F) at Worcester
4.	19 August	1932	36.1 °C (97.0 °F) at several locations in the London area and Halstead (Essex)
5.	3 July	1976	35.9 °C (96.6 °F) at Cheltenham
6.	2 July	1976	35.7 °C (96.2 °F) at Cheltenham
7.	28 June	1976	35.6 °C (96.1 °F) at Southampton
8.	2 September	1906	35.6 °C (96.0 °F) at Bawtry (nr. Doncaster)
	13 August	1911	35.6 °C (96.0 °F) at Salisbury
	13 July	1923	35.6 °C (96.0 °F) at Camden Sq. (London)
	29 June	1957	35.6 °C (96.0 °F) at Camden Sq. (London)
12.	4 August	1990	35.5 °C (95.9 °F) at Kew Gardens
13.	27 June	1976	35.5 °C (95.9 °F) at Southampton
14.	26 June	1976	35.4 °C (95.7 °F) at North Heath (West Sussex) and East Dereham (Norfolk)
15.	20 July	1900	35.2 °C (95.4 °F) at Cambridge
16.	1 July	1952	35.1 °C (95.2 °F) at St Helier (Jersey)

17.	1 September	1906	35.0 °C (95.0 °F) at Collyweston (nr. Stamford), New Malden and Maidenhead
	8 August	1911	35.0 °C (95.0 °F) at Kingston-on-Soar (Notts)
	13 July	1923	35.0 °C (95.0 °F) at Bristol, Reading, and Hitchin
	18 August	1932	35.0 °C (95.0 °F) at St Helier (Jersey)
	28 July	1948	35.0 °C (95.0 °F) at Milford (nr. Godalming)

THE COLDEST NIGHTS OF THE TWENTIETH CENTURY

1.	10 January	1982	−27.2 °C (−17.0 °F) at Braemar
2.	8 January	1982	−26.8 °C (−16.2 °F) at Grantown-on-Spey
3.	11 January	1982	−26.6 °C (−15.9 °F) at Bowhill (Selkirk)
4.	13 December	1981	−25.2 °C (−13.4 °F) at Shawbury (Shropshire)
4.	23 February	1955	−25.0 °C (−13.0 °F) at Braemar
5.	13 January	1979	−24.6 °C (−12.3 °F) at Carnwarth (Lanark)
6.	20 January	1984	−23.6 °C (−10.5 °F) at Grantown-on-Spey
7.	28 January	1910	−23.3 °C (−10.0 °F) at Balmoral and Logie Coldstone
	14 November	1919	−23.3 °C (−10.0 °F) at Braemar
	28 January	1940	−23.3 °C (−10.0 °F) at Rhayader
10.	14 March	1958	−22.8 °C (−9.0 °F) at Logie Coldstone

THE 15 SUNNIEST TOWNS IN THE BRITISH ISLES
(based on 30-year averages)

1.	St Helier (Jersey)	1917 hours per year
2.	St Peter Port (Guernsey)	1892 hours
3.	Shanklin (Isle of Wight)	1887 hours
4.	Sandown (Isle of Wight)	1858 hours
5.	Ventnor (Isle of Wight)	1839 hours
6.	Worthing (Sussex)	1821 hours
7.	Eastbourne (Sussex)	1810 hours
8.	Hastings (Sussex)	1773 hours
	Littlehampton (Sussex)	1773 hours
	Hayling Island (Hants)	1773 hours
11.	Weymouth (Dorset)	1771 hours
12.	Ryde (Isle of Wight)	1766 hours
13.	Folkestone (Kent)	1760 hours
	Bognor Regis (Sussex)	1760 hours
	Swanage (Dorset)	1760 hours

THE 13 DRIEST TOWNS IN GREAT BRITAIN
(based on 30-year averages)

1.	Grays Thurrock (Essex)	510 mm (20.08 in)
2.	Sheerness (Kent)	530 mm (20.87 in)
3.	Felixstowe (Suffolk)	540 mm (21.26 in)
4.	Dagenham (Gtr London)	550 mm (21.65 in)
	Tilbury (Essex)	550 mm (21.65 in)
	Southend (Essex)	550 mm (21.65 in)
	Colchester (Essex)	550 mm (21.65 in)
	Ipswich (Suffolk)	550 mm (21.65 in)
	Cambridge (Cambs)	550 mm (21.65 in)
	Ely (Cambs)	550 mm (21.65 in)
11.	Bedford (Beds)	560 mm (22.05 in)

Peterborough (Cambs)	560 mm (22.05 in)
Clacton (Essex)	560 mm (22.05 in)

THE 12 WETTEST TOWNS IN GREAT BRITAIN
(based on 30-year averages)

1.	Fort William (Inverness)	2010 mm (79.13 in)
2.	Blaenau Ffestiniog (Merioneth)	2000 mm (78.74 in)
3.	Ambleside (Westmorland)	1851 mm (72.87 in)
4.	Maesteg (Glamorgan)	1800 mm (70.87 in)
	Dolgellau (Merioneth)	1800 mm (70.87 in)
6.	Aberdare (Glamorgan)	1750 mm (68.90 in)
	Rhondda (Glamorgan)	1750 mm (68.90 in)
8.	Tredegar (Gwent)	1600 mm (63.00 in)
	Mountain Ash (Glamorgan)	1600 mm (63.00 in)
	Windermere (Westmorland)	1600 mm (63.00 in)
11.	Greenock (Renfrew)	1570 mm (61.80 in)
	Pontypridd (Glamorgan)	1570 mm (61.80 in)

THE 10 WETTEST MONTHS SINCE 1766
(over England and Wales)

1.	October	1903	218 mm (8.59 in)
2.	November	1852	203 mm (7.97 in)
3.	November	1770	201 mm (7.91 in)
4.	November	1940	197 mm (7.74 in)
5.	November	1929	196 mm (7.71 in)
6.	December	1876	194 mm (7.63 in)
7.	August	1912	193 mm (7.59 in)
8.	August	1799	192 mm (7.56 in)
9.	December	1914	191 mm (7.51 in)
10.	September	1918	189 mm (7.46 in)

THE 10 DRIEST MONTHS SINCE 1766
(over England and Wales)

1.	February	1891	3.6 mm (0.14 in)
2.	June	1925	4.3 mm (0.17 in)
3.	January	1766	4.4 mm (0.18 in)
4.	March	1781	5.6 mm (0.22 in)
5.	April	1938	7.1 mm (0.28 in)
6.	April	1817	7.9 mm (0.31 in)
	May	1844	7.9 mm (0.31 in)
8.	March	1929	8.0 mm (0.32 in)
	September	1959	8.0 mm (0.32 in)
10.	July	1825	8.2 mm (0.33 in)

GENERAL INDEX

PLACE-NAME INDEX